EARLY AMERICAN STEAM LOCOMOTIVES

REED KINERT

Roger M. Clemons
P. O. Box 102
Bartlett, NH 03812

DOVER PUBLICATIONS, INC.
Mineola, New York

Bibliographical Note

This Dover edition, first published in 2005, is an unabridged republication of the work originally published by Superior Publishing Company, Seattle, Washington, in 1962 under the title *Early American Steam Locomotives, 1ˢᵗ Seven Decades . . . 1830–1900*.

Library of Congress Cataloging-in-Publication Data

Kinert, Reed.
 Early American steam locomotives / Reed Kinert.
  p. cm.
 Reprint. Originally published: Seattle, WA : Superior Pub., 1962.
 Includes index.
 ISBN 0-486-44398-1 (pbk.)
  1. Steam locomotives—United States—History—19th century. 2. Railroads—United States—History. I. Title.

TJ603.2K55 2005
625.26'1'0973—dc22

                                                    2005049621

Manufactured in the United States of America
Dover Publications, Inc., 31 East 2nd Street, Mineola, N.Y. 11501

AUTHOR'S NOTE

It was the clanking, puffing, belching and sooty "Iron Horse" that welded our vast country into one nation, cutting travel time coast to coast from weeks and months into days.

Invented in England and first used successfully there in 1825, the steam locomotive concept soon found its way to America, where it was quickly adapted to our needs. Now, with the coming of motor trucks to the highways and the airliners to the skyways, steam power as a means of locomotion has suddenly all but disappeared from many of America's rails while the oil burning diesel-electric engine takes over to more economically compete with the new arrivals.

While the diesel locomotive is an undebatable improvement in the efficiency of the machines themselves, the interest of the machine seems to have disappeared into a sheath of smooth metal. Even the sound is as different as night and day—the staccato bark of the steamer's exhaust and her eerie whistle giving way to the smooth throb of diesel engines and a long honnk honnk, this latter sound being likened to the mating call of a love-sick moose!

From first to last the steam engine was a "contraption," an assembly of parts seemingly stuck together with rods, braces and bolts. The engine has always been a hybrid assembly of a cart, a furnace and a prime mover. In an uneasy combination it has had to reconcile its undercarriage of frame and wheels, its complex mechanism of cylinders and rods and its unwieldy steam generating plant. Although these, plus a house for the engineer, have always made an awkward union, the very difficulty of uniting them has insured the vitality of shape that resulted—a fascinating hunk of machinery that, when fired-up, pulsed and throbbed like a live monster eager to be on its way.

While the English from the very first built their locomotives on a smooth cart-like rigid frame that passed outside the wheels and dominated the design, the American designer, with an inventive skill that must have come from frontier training, put the frame inside the wheels, where it was all but invisible. There is a feeling that our machines might be assembled with scraps of wire and string and it is a fact they actually do use the track they run on to hold their parts together!

It is a paradox that this crude and rough appearing frontier locomotive was actually a more advanced mechanism than its polished English cousin. Eliminating the cart-like frame saved weight and material, and gave a flexible machine that was far better adapted to the uncertainties of early railroad track and to limited repair facilities.

American steam locomotives took many forms and varied in appearance from cute to brute, from homely and even ugly to handsome or beautiful. In these pages are depicted many famous and colorful early engines in all their glory, engines that show a growth from the "teakettle" class to the really practical types.

REED KINERT

LIST OF ILLUSTRATIONS

Single asterisk* denotes color illustration   Double asterisk** denotes photograph

	Page		Page		Page
A. J. Cromwell	116**	General	90**	Tom Thumb	11, 33*
Alert	37*	General Darcy	138	Washington	38*
Andrew Jackson	85**	Genoa	106**	West Point	13, 26
Anthracite	139	Governor Williamson	131	William Crooks	92**
Arabian	84**	Gowan & Marx	55	William Mason	7, 44*, 91**, 141, 159
Asa H. Curtis	112**	Hercules	51	Willie	110**
Atlantic	35*, 52	Indiana	40*	York	14
Baldwin Flexible Beam truck	80	J. C. Davis	100**	0-4-0 type, Porter	118**
Baldwin 6-wheel connected	59	John Hancock	31, 86**	0-4-0T type, No. 7	103**
Best Friend of Charleston	20, 34*, 81**	John Stevens	75, 76	0-4-0T type, No. 569	104**
Black Hawk	74	Lafayette	36*	0-4-2T type, No. 7	97**
Camel No. 217	48*	Lightning	73	2-4-4TB suburban type	102**
Camel type	146	J. W. Bowker	99**	2-6-0 type, No. 3	120**
Campbell American type	30	Memnon	87**	2-6-0 type, No. 4	121**
C. L. Flint	145	Minnetonka	96**	2-6-4TB suburban type No. 1447	119**
Carbon	135	Mississippi	83**	2-8-0 type, No. 229	149
Chesapeake	70	Mogul type, first	69	2-8-0 type, No. 667	109**
Cincinnati	79	Monster	28	4-4-0 American type, No. 1	101**
Chloe	108**	Mud Digger, first type	67, 80	4-4-0 American type, No. 3	113**
Claus Spreckels	128**	Mud Digger, second type	68	4-4-0 American type, No. 3	115**
Climax geared type No. 770	122**	Norris six-coupled	63	4-4-0 American type, No. 11	58
Coach, Winans double truck	64	Old Ironsides	22	American type, No. 119	123**
C. P. Huntington	45*	Olomana	107**	American type, No. 137	47*
Crab type 0-4-0	52, 54	Perry	39*	American type, No. 140	127**
Daniel Nason	89**	Phoenix	21	American type, No. 173	98**
DeWitt Clinton	16, 17, 82**	Pioneer	41*, 88**	American type, No. 374	114**
Dorothy	111**	Rail types	12	American type, No. 870	151
Dutch Wagon type	89**, 132	Reuben Wells	94**	American type, No. 999	155
E. L. Miller	27	Smoke stack types	140	4-6-0 type, No. 70	125**
Emma Nevada	105**	South Carolina	19	4-6-0 type, No. 2174	117**
Experiment	25	Tank type 0-6-0	42*	4-4-2T type, No. 1	95**
Forney type, first suburban	142	Thatcher Perkins	46*, 93**	4-8-0 type, No. 2933	124**
Forney type, No. 2	126**	Tiger	43*		

TABLE OF CONTENTS

Author's Note - - - - - - - - - - - - - - - - 5

Chapter I   America's First Steam Locomotive - - - - - 9

Chapter II   America's Second and Third Steam Locomotives - 15

Chapter III   The Iron Horse Begins to Take Over - - - - 23

Chapter IV   Another Watchmaker Appears - - - - - - 29

Chapter V   During Which Steady Progress Is Made - - - 49

Chapter VI   The "Grasshoppers" Have Their Day - - - - 53

Chapter VII   The Wondrous Steam Engine Grows - - - - 57

Chapter VIII   Canals, Roads and Railroads - - - - - - 61

Chapter IX   Bells, Whistles, Headlights and Other "Extras" - 65

Chapter X   New Rails and Larger Engines - - - - - - 71

Chapter XI   Mud Diggers, Camels and High Drivers - - - 77

Chapter XII   The Wild Iron Horse - - - - - - - - - 129

Chapter XIII   The American Type Locomotive - - - - - - 133

Chapter XIV   The Glorious Iron Horse - - - - - - - - 137

Chapter XV   Civil War and Rails West - - - - - - - 143

Chapter XVI   Cylinders, Boilers, Coal and Huge Engines - - 147

Chapter XVII   Eulogy of Steam - - - - - - - - - - - 153

Specification Chart - - - - - - - - - - - - - 156

Index - - - - - - - - - - - - - - - - - 157, 158

ACKNOWLEDGEMENTS

I should like to thank all the railroads of the United States and Canada, and the American Association of Railroads for their invaluable aid in preparing this book. Special mention is deserving of the following: Mr. Lawrence W. Sagle of the Baltimore & Ohio; the Motive Power Department, Mr. Charles R. Dunlap, and Manager of Publicity, Mr. G. E. Payne, all of the Pennsylvania Railroad. Also a special thanks to Mr. A. C. Carson, Manager of Press Relations, Illinois Central Railroad, and Mr. B. E. Young, Assistant to the President, Southern Railway System. And last but not least, Mr. J. G. Shea, General Public Relations Manager, Southern Pacific Company.

My thanks extends greatly to Ward Kimball for his valuable assistance and to Gerald M. Best for his outstanding help in the photo section of our book.

REED C. KINERT

CHAPTER I

AMERICA'S FIRST STEAM LOCOMOTIVE

The first American railroads were powered by horses, just as were the wagons upon the roads and the boats on the canals, and freight and passengers traveled over all three forms of transportation at the same leisurely pace of four miles an hour. But railroads, from the very first, held decided advantages over roads and canals. A horse could, over rails, pull 20 to 30 times more load than he could over a dirt road and while a canal and railroad horse could pull about the same tonnage of freight, the railroad cost a lot less to build and maintain per mile than the canal. The railroad could go anywhere, not being restricted to rather level land like the canals were.

The Baltimore and Ohio Railroad was the first American railway public carrier in regular service. Chartered on February 28, 1827 with plans to connect Baltimore with the boat traffic of the Ohio River some 379 miles westward, at Wheeling, West Virginia, construction began in late 1828 with iron strap rails being doweled or screwed to wood stringers, the stringers then being set in notched ties which were set into a gravel roadbed — and the work was done by carpenters!

While work progressed steadily, it was May 24, 1830 before the horse drawn cars were able to pull into Ellicott's Mills, only thirteen miles westward. Meanwhile Baltimore business men and the railroad directors had been meeting often, for they grew increasingly worried about the progress of new railroads and canals to the north and south of them. Into such a directors meeting in the fall of 1829 strode Mr. Peter Cooper, a most amazing merchant from New York City who had invested in the lands of a terminal and improvement company in Baltimore.

Mr. Cooper too was worried about his investment in Baltimore, as its success hinged very heavily upon the success of the town itself. Mr. Cooper told the directors that he believed steam locomotion might be the answer to their problems and that he believed he could knock together a workable steam locomotive for testing on the rails.

The world's first practical steam locomotive for use on rails had been built in England by George Stephenson in 1825. Placed on the rails of a horse operated railroad, Stephenson's engine named *Locomotion*, with George as engineer, pulled twenty-nine small four-wheeled cars plus a tender loaded with water and coal, and averaged eight miles per hour. Then, early in 1829, four English locomotives were purchased and shipped to northeastern Pensylvania to be used by the Delaware and Hudson Canal Company in hauling coal from it's Carbondale mines to the canal terminus, sixteen and a half miles east at Honesdale. Upon arrival the locomotives were found to weigh almost seven tons instead of their estimated three tons. Two of the four engines were assembled at the West Point Foundry, New York City, and one of them, the *Stourbridge Lion*, was tried in August of 1829 on the Delaware & Hudson's rails but was found to be too rigid for the uneven and sharply curved track in addition to being just too heavy for the flimsily built roadbed. All four locomotives eventually disappeared, the other three apparently never being set on rails. Then in 1905 the boiler and other parts of the celebrated *Stourbridge Lion* were collected in various parts, sent to the Smithsonian Institute, Washington, D.C., and put together, making, with the addition of a few new pieces, a complete engine. The Baltimore & Ohio directors had investigated the above tests so figured they had nothing to lose by giving Mr. Cooper a chance at building a locomotive. Peter, who owned a foundry and was handy with tools, journeyed to New York and bought a little steam engine with a cylinder three and one-quarter inches by fourteen and one-half inches and, returning to Baltimore, got some iron and built a boiler about as big as an ordinary wash boiler of that day. He wanted some iron pipes for boiler flues but none were obtainable so he used two old musket barrels. Peter went into a coachmakers shop and made his locomotive which he called Tom Thumb because it was so small. He didn't intend it for actual service but only to show the directors what could be done. He meant to show two things: first, that short turns could be made; and, secondly, that he could get rotary motion without use of a crank. He changed the engines reciprocating movement to a rotary motion by using two gears. To emphasize the fact that the Tom Thumb was developed entirely independent of the English we note the prior invention by Mr. Cooper of two fundamental features later used on all steam locomotives—the multi-tubular boiler and the artificial draft. Cooper's boiler tubes of musket barrels have been mentioned. The air blast by which he secured a forced draft was obtained by a blower fan worked by a belt from the axle of the engine. Then, later, Stephenson hit upon the correct principle of a forced draft, quite by accident. The exhaust from his first locomotive frightened horses and he was notified by the police that if he kept on with his noisy engines he would be arrested. He turned the exhaust into the smoke-stack to muffle the noise and found to his joy that he had provided a forced draft. It is this latter principle that is in use to this day while Peter Cooper's blower idea wasn't used until much later when adapted to supercharge reciprocating gasoline and diesel engines.

Finally the little locomotive was finished, then one Saturday night steam was got up; the president of the road and two or three gentlemen had been anxiously standing by so they and Peter climbed onto Tom Thumb and went out two or three miles towards the west. All were very much delighted and Tom Thumb was put in its shed and all were invited to a daytime ride on Monday—a ride to Ellicott's Mills. Came Monday and what was their grief and chagrin to find that some scamp had been there and chopped off all the copper from the engine and carried it away. The copper lines that conveyed the steam to the piston was gone. It took Mr. Cooper a week or more to repair the locomotive, which, you may be sure, was placed under guard each night, and then on Saturday, August 28, 1830, with six men on the engine and thirty-six men in an attached car, the Tom Thumb began the first trip by an American-built locomotive. It went up an average grade of eighteen feet to the mile and made the passage to Ellicott's Mills in an hour and twelve minutes (this included watering and track switching time) and averaged five and one-half miles per hour.

Then, on the return trip from Ellicott's Mills, an event took place that has been made much of in the history of railroading. It was an interesting trip. The shortest of curves were traversed without materially slackening a speed of fifteen miles an hour. The day was fine, the company in the highest spirits and some excited gentlemen of the party, when the train was at its highest speed, eighteen miles an hour, pulled out their note books and wrote connected sentences just to show that such a thing was possible!

Then Tom Thumb and its gay load of passengers arrived at Relay House, the halfway home point, where it was known the little locomotive would be refueled. The great stage proprietors of the day, Stockton and Stokes, who furnished most of the horses used by the B & O, had contrived to have a gallant grey horse of great beauty and power hitched to an enclosed passenger car standing ready on the second track—for the company had wisely begun by making two tracks to the mills.

Fearful of the steam locomotives eventual success, but still confident of the horses superiority over the engine, Stockton &

The little TOM THUMB slowly gained on the gallant grey horse then passed him and it seemed that the race was won.

Stokes had arranged this meeting to prove their point. They and their cohorts jeered and coaxed for a race and Peter Cooper was not unwilling for he felt his little engine to be the master of any horse.

The start being even away went horse and engine, the snort of one and the puff of the other keeping time and tune. At first the grey had the best of it, for his power was applied to the greatest advantage on the instant, while the engine had to wait until rotation of the wheels set the blower to work. The horse was perhaps a quarter of a mile ahead when the safety-valve of the engine lifted and the thin blue vapor issuing from it showed an excess of steam. The blower whistled, the steam blew off in vapory clouds, the pace increased, the passengers shouted. Soon the race was neck and neck, then the engine passed the horse and a great hurrah hailed the victory. But it was not repeated; for just at this time, when the grey's master was about to give up, the band which drove the pulley which drove the forced-draft blower, slipped from the drum. The safety-valve ceased to scream and the engine, for want of breath, began to wheeze and pant. In vain Mr. Cooper, who was his own engine man and fireman, lacerated his hands in attempting to replace the band upon the wheel; in vain he tried to urge the fire with light wood. The horse gained on the machine and passed it; and although the band was presently replaced and steam again did its best, the horse was too far ahead to be overtaken and so came in the winner.

The victory of the grey horse was short lived and proved to be the last gasp of a mode of transport that was about to expire upon the railroad. The horse and the order of things for which it stood was doomed. Peter Cooper soon fixed the fan belt upon the blower so it wouldn't slip off and presently was out on the track again. For some weeks the Tom Thumb continued to operate on the railroad, giving demonstrations, and important data was being compiled on the little locomotive which could pull a load of forty-two people up a slight grade with its tiny engine developing but 1.43 horsepower!

THE EVOLUTION OF RAIL

EARLY TYPES

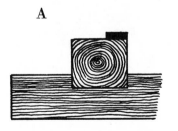
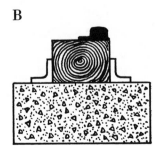

A. First strap rail—rolled iron.
B. Another form of strap rail.

C   D   E   F   G

C. Inverted "U" shaped iron rail.
D. Inverted "U" rail closed.
E. Pear shaped iron "T" rail.
F. 50 lb/yd steel rail.
G. Modern 132 lb/yd plus high carbon steel "T" rail.

THE WEST POINT

2ND LOCOMOTIVE OF THE SOUTH CAROLINA CANAL & RAILROAD CO.

Builder . . . West Point Foundry, New York City . . . 1831

THE YORK

4 Wheel Vertical-Boilered Engine

Builder . . . Davis & Gartner . . . 1831

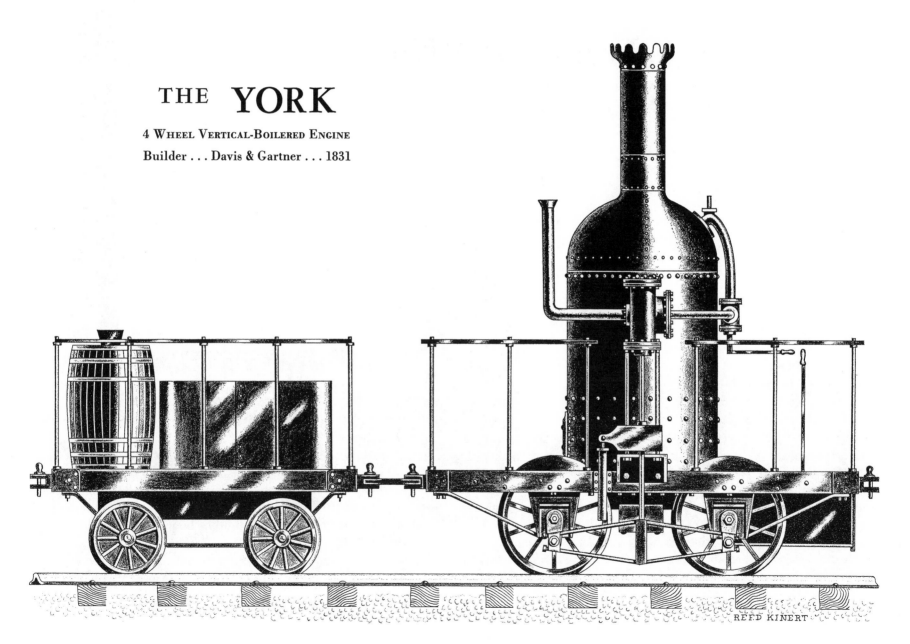

CHAPTER II

AMERICA'S SECOND & THIRD STEAM LOCOMOTIVES

While the Baltimore & Ohio Railroad holds the honor of being the first American railway carrier in regular service, steam did not completely displace horse-drawn traffic over that road until several years later. By contrast, the South Carolina Canal & Railroad Company (now part of the Southern Railway) used steam from its beginning. The railway was to be built from Charleston, S.C., thence west northwest to Hamburg, S.C., just across the Savannah River from Augusta, Georgia, the distance of 136 miles.

Work began on the railroad at Charleston January 9, 1830, and iron-strap rails were spiked to wood stringers, the stringers set into wooden cross ties in much the same type construction as used on the B & O.

With the railroad under way, their new chief engineer, Horatio Allen, who had acted as purchasing agent in buying and assisted in assembling of the English locomotives mentioned in chapter I, convinced the railroad directors they should use steam locomotion from the start. With the help of C. E. Detmold, who had invented for the railroad a horse-driven motor worked on an endless chain platform, and E. L. Miller of Charleston, the three men proceeded to design a unique engine. The West Point Foundry of New York was commissioned to build the locomotive since they had gained some experience by assembling the two imported English engines mentioned, under E. L. Miller's supervision. The foundry built the locomotive which was wisely named *Best Friend of Charleston*, for $4,000 and on October 23, 1830, it arrived at Charleston aboard the steamship Niagara. Nearly all the townfolk eagerly turned out to get a look at the steam horse which they hoped

would make Charleston a great seaport by providing low cost transportation between their port and the fast developing inland areas.

Best Friend was a lightweight, weighing less than four tons and it developed only six horsepower. All four wheels were drivers, motivated by a double crank from two inclined cylinders mounted inside and at the front of the frame, each wheel being connected together with outside rods. The wheels were iron hub, the spokes being made of hard wood and the wheel tires were iron. The boiler was a vertical one, in the form of an old-fashioned porter bottle and looked somewhat like a Coke bottle too. The *Best Friend* made several short trips upon the rails, then on November 2, 1830, with her engineer Nicholas W. Darrell in charge, a trial trip was made hauling several men in an attached car. In rounding a curve the forward wheels left the rails and the wood-spoked front wheels all but collapsed as the engine rolled to a stop some 20 feet ahead. The two crewmen escaped the accident with some bruises; the men in the following car were uninjured.

More than a month's repairing was necessary before *Best Friend* was reworked with iron-spoked wheels and made ready for another unofficial jaunt. On December 14, 1830, the engine proved successful by pulling two "coaches" of 14 foot length and which were fitted with a double bench running down the middle—a sort of wooden "pew." Forty not too willing passenger recruits from among the company's laborers were hauled at a top speed of twenty miles an hour on one stretch.

On Christmas Day of 1830 the *Best Friend of Charleston*, hooked to its two high-sided and roofed wooden wagons with

THE DeWITT CLINTON

NEW YORK STATE'S 1ST STEAM LOCOMOTIVE

Builder . . . West Point Foundry . . . 1831

The DeWITT CLINTON, New York State's first steam locomotive, was set onto iron strap-top wooden rails that had seen only horse-drawn cars. Note how the wood rails (stringers) were set into the wooden cross-ties.

The original concept of the DEWITT CLINTON is shown at top while below is a silhouette cut out of black paper with scissors by a singularly gifted man named W. H. Brown, who made the silhouette at the scene of the CLINTON's first excursion trip August 9, 1831. Note the added steam dome.

bare board seats, headed out of Charleston with a load of paying passengers to traverse the then completed first six miles of railroad and thus became America's first regularly scheduled steam railroad carrier. One passenger describing the trip stated, "Away we flew on the wings of the wind at the speed of fifteen to twenty-five miles an hour, scattering sparks and flames on either side, passed over three saltwater creeks, hop, step and jump and reached the end of the line at State and Dorchester Roads before any of us had time to determine whether or not it was prudent to be scared."

The *Best Friend* worked at hauling materials for construction of the new railroad as well as making passenger excursion runs on the ever westward building railroad. Then, on June 17, 1831, another accident befell the *Friend*. A negro fireman, annoyed by the constant hissing of escaping steam from the safety valve, fastened the valve lever down, the excessive pressure fractured the crown sheet and the reaction threw the boiler in the air. Engineer Nicholas Darrell was busily directing arrangement of lumber cars on the turn table at Forks of the Roads, was scalded in the ensuing boiler explosion, which scalded the fireman so badly he later died and the *Best Friend* was blasted out of business. As a result of this accident the S.C.C. & R.R. Co. thereafter placed a barrier car between locomotive tender and the passenger cars, then strapped six bales of cotton to it for the passengers' protection from steam or hot water, should a boiler break occur—a not too infrequent occurrence during railroading's early days.

Before the *Best Friend* boiler exploded the S.C.C. & R.R. Co. had gone ahead with their plans for an all steam railroad. Horatio Allen had drawn up a set of plans for their second engine, America's third home built locomotive and the first engine built in the U.S. to have a horizontal boiler. The boiler had a square "wagon top" firebox with no dome on top. Six or eight tubes about 3″ in diameter and 6′ long conveyed the com-

bustion heat from the firebox through the boiler to the forward smoke box, which was merely the base of the smokestack. Named the *West Point* after the foundry in which it was built, the little four wheeler arrived at Charleston aboard the ship Lafayette and was put on the track with Darrell again as the engineer. Like the *Best Friend* and all other early locomotives on the S.C.C. & R.R. Co., the *West Point* was fired with rich pitch-pine wood, known in the South as "lightwood," a fuel which produced such volumes of dense black smoke that, however picturesque, it must have been most unpleasant for passengers riding behind the smokestack in the box-like open sided "coaches."

The *Charleston Courier* newspaper of March 12, 1831, related that "On Saturday afternoon March 5, 1831, the locomotive *West Point* underwent a trial of speed, with the barrier car and four cars for passengers. There were 117 passengers aboard, of which 50 were ladies, in the four cars and 9 passengers on the engine tender, with 6 bales of cotton on the barrier car. The trip to 5 Mile House, two and three-quarter miles, was completed in 11 minutes, where the cars were stopped to oil the axles about 2 minutes. The two and one-quarter miles to the forks of Dorchester Road were completed in 8 minutes." The new locomotive worked admirably and was soon put to work hauling trains on a regular schedule. After the explosion of the *Best Friend*, the safety valve was re-located so it was out of reach of any person but the engineer to prevent such accidents.

The Southern Railway has built an exact and beautiful replica of the original *Best Friend of Charleston* which differs from the original only in that the wheels now fit standard gauge track while the *Best Friend* rode over track set 5 feet apart. It is interesting how the gauge of track came to be standardized, or nearly so, over the United States and England. Pioneer locomotiveman George Stephenson learned that the distance

THE SOUTH CAROLINA

1st Articulated Locomotive

Builder . . . West Point Foundry . . . 1832

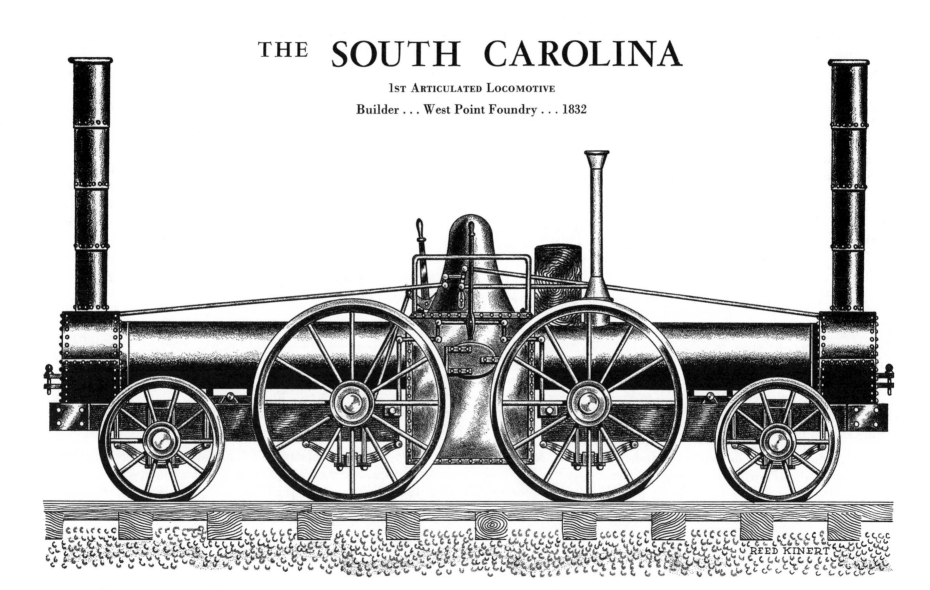

between the wheels of all vehicles, from the earliest of Roman chariots to the latest carriages and coaches of his native England seemed to average out at 4 feet eight and one-half inches, so he decided to build his locomotives with their wheels this distance between them. This seemingly magic measurement was found on the engines shipped to this country so the Yankee designers accepted the figure without question. There were quite a few variations from this width at first but the railroads that had not conformed were put to no little expense to change their gauge. An exception to this standard gauge is the so named narrow gauge, all being several inches less than standard, some going down to as little as two feet. There are still some narrow gauge railroads to be found in this country, but the number grows ever less.

It was passenger and not freight traffic that kept the early railroads running. A case in point and striking proof that railroads actually created travel and thus growth to themselves is to be seen in the experience of the towns of Charleston and Hamburg in South Carolina. Before the railroad was built, passenger traffic between the two towns had been adequately cared for by one stagecoach, which carried only six passengers without undue crowding and which made only three trips a week. During six months of operation in 1835 the railroad carried 15,959 passengers, the comparison in numbers per month being 50 travelers on the stage coaches as against 2,500 on the steam trains.

A most fortunate thing about these primeval railroads was that, crude as their tracks and rolling stock were, there were few great disasters due to accidents. The average speed was probably not more than 18 mph and there was little travel by night. The idea of close time schedules and the constant desire of engineers to "make up time" and "bring her in on the advertised" was not present. Departures were usually advertised and pretty well adhered to but from then on it was a game between man and machine to arrive at their destination within any reasonable estimate. Certainly travel by rail would not have increased so rapidly as it did had there been many wrecks disastrous to human life—the kind of wrecks that, beginning in the 1850's were a national scandal.

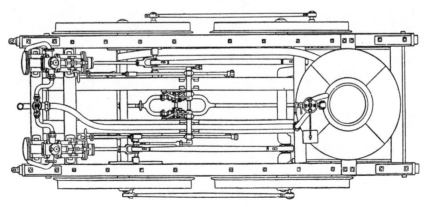

Top scale drawing of the *Best Friend of Charleston*

THE PHOENIX

4 Wheel Type Locomotive

Rebuilt from the *Best Friend of Charleston* engine . . . 1832

OLD IRONSIDES

Matthias Baldwin's 1st Steam Locomotive

Builder . . . Matthias W. Baldwin & Co., Philadelphia . . . 1832

CHAPTER III

THE IRON HORSE BEGINS TO TAKE OVER

In its beginning the Baltimore and Ohio Railroad rented its horses from the stagecoach companies and no horse was required to pull cars (named wagons) more than six or seven miles. Including food, stabling and necessary personnel to groom and care for the horses, accounting showed that work done by horsepower cost $33.00 a day per train, or brigade as they were called in those days. This, plus the difficulties B & O management foresaw in using horsepower all the way to the Ohio River, with the tremendous number of relay stable stations to equip and maintain, caused them to envision complete replacement of horses by steam power.

In the first week of January 1831 the B & O directors, convinced by the showing of *Tom Thumb* and the success of English locomotives abroad that steam could combine economy with speed, offered a prize of $4,000 for the best steam engine and $3,000 for the second best steam engine to be entered in a contest scheduled for June of that year and specifications were placed in Baltimore, Pittsburgh, New York and Philadelphia newspapers in detail. They called for a coal or coke burning engine that would weigh less than three and a half tons and be capable of drawing, "day by day," fifteen tons, inclusive of the wagons, fifteen miles per hour. "Their boilers must not exceed 100 lbs. pressure to the square inch, lower pressure preferred, and there must be *two* safety valves, one of which must be completely out of control of the engine men and neither of which must be fastened down while the engine is working." Other details were specified and the railroad required thirty days trial before acceptance. The company would provide a tender, fuel and water.

Five engines entered the contest, which was won by the *York*, designed and built by Phineas Davis, a watchmaker from the Pennsylvania town for which his engine was named. Davis had been keenly interested in locomotives, studying everything he could find about their construction and operation. When he heard of the B & O contest for a locomotive, Davis formed a partnership with a machinist named Gartner and they had completed the *York* in only a few months. Davis' *York* not only met the B & O specifications, but negotiated the sharpest curves at the maximum fifteen mph required and made bursts of thirty-five mph on the straightaways. Further, it was found that the *York* could be operated for $16.00 a day, less than half as much as the average for animal powered trains.

Mr. Davis' *York* was shorter in length than *Tom Thumb*, but it was vastly superior, partially due to its having two cylinders for power while *Tom Thumb* had but one. The *York* was, like the *Tom Thumb*, mounted on four wheels, but all four were used for tractive effort. With the *York*, Phineas Davis not only won his $4,000 but a position as Master Mechanic of the B & O.

In July of 1831 the *York* was placed on a one-trip-a-day schedule on the run between Baltimore and Ellicott's Mills, hauling as many as five passenger cars in a brigade and later on the much longer run of some 40 miles between Baltimore and the inclined planes at Parr's Ridge, on the way to Fredericktown, Maryland. Successful as the *York* proved to be, it was regarded as only a forerunner of better engines to come.

As a result of the locomotive trials the B & O railroad became known as the railroad university for they made their findings on locomotives freely available to all those interested and also released general information on the building and running of their railroad.

The success of the B & O trials probably did more than any one thing to spur the building of other U.S. railroads and began naturally a rash of locomotive building. Everyone wanted to build a railroad. They didn't seem to care where they built one, just so it ran between two towns. By 1835 more than 200 railroad charters were granted in eleven states, but only a few of these early chartered companies ever succeeded in actually completing railroads, and only a few of those reaching completion operated for more than a brief period of time. As evidence of this is the fact that at the end of 1835 there were only 1,098 miles of rails laid and in operation in the U.S.; still a good start.

The year 1830 had seen America's first two locomotives, *Tom Thumb* and the *Best Friend of Charleston*. In 1831 three locomotives were built in the U.S., and two of them were constructed by the West Point Foundry, which continued its pace by building the *West Point* locomotive for the South Carolina Canal & Railroad Company, then later in the year, their now famous *DeWitt Clinton* engine for the new Mohawk & Hudson Railroad (now part of the New York Central) which began by running horse-drawn cars between Albany and Schenectady, N. Y., a rail distance of only 15.9 miles—but it was New York State's first railroad.

The *DeWitt Clinton* engine was built by David Matthews, a mechanic at the foundry, from plans drawn up by John B. Jervis, Chief Engineer of the Mohawk & Hudson Railroad.

The completed *Clinton* was shipped up the Hudson River on June 25, 1831, and a week later Matthews had steam up—then his troubles began. In the first place it was found that the chimney (stack) was too large, and the ends of the exhaust pipes—they hadn't got around to nozzles yet—were placed so low they couldn't get up a draft; so some rebuilding had to be done at once.

When the *Clinton* was given a second trial the water in its boiler surged over the cylinders, threatening to knock them to pieces, as the engine bobbed over the rough and uneven track. A steam dome was built and installed atop the boiler and that trouble was overcome.

Finally, on August 9, 1831, the *Clinton* was deemed ready for its first scheduled run from the outskirts of Albany to Schenectady and five coaches were provided for passengers. The cars were literally coaches—stage coaches mounted on flanged wheels and coupled together with three long links of chain between each two coaches. The occasion was one of great public rejoicing and several hundred brave people failed to buy seat space for lack of accommodation.

The conductor, a Mr. Clark, took his seat at the rear of the tender and blew his trumpet for the starting of the train, the first scheduled steam train in the state of New York. Matthews opened the throttle with a suddenness which almost snapped the passengers' heads off, fashionable hats flew in all directions and a few passengers with them, all because of the great amount of slack between coaches. As the *Clinton* moved on at an increasing speed the jerking finally tapered off into a steady pull as the cheers of the admiring spectators died away in the distance. By the time the passengers drew a long breath and looked about them, they found new tribulations.

The *DeWitt Clinton* had been built for anthracite coal, but the coal packed and wouldn't burn until a steam-blower was put on, then the grates burned out, so they had had to use pitch-pine as the fuel for this first trip. The result was that a steady stream of pitch-pine cinders, from the size of a pin-head to that of a man's thumb-nail poured back on the passengers. Umbrellas were raised for protection but soon burned up like so much

THE EXPERIMENT

WORLD SPEED RECORD: 80 M.P.H. (UNOFFICIAL)

Builder . . . West Point Foundry, New York City . . . 1832

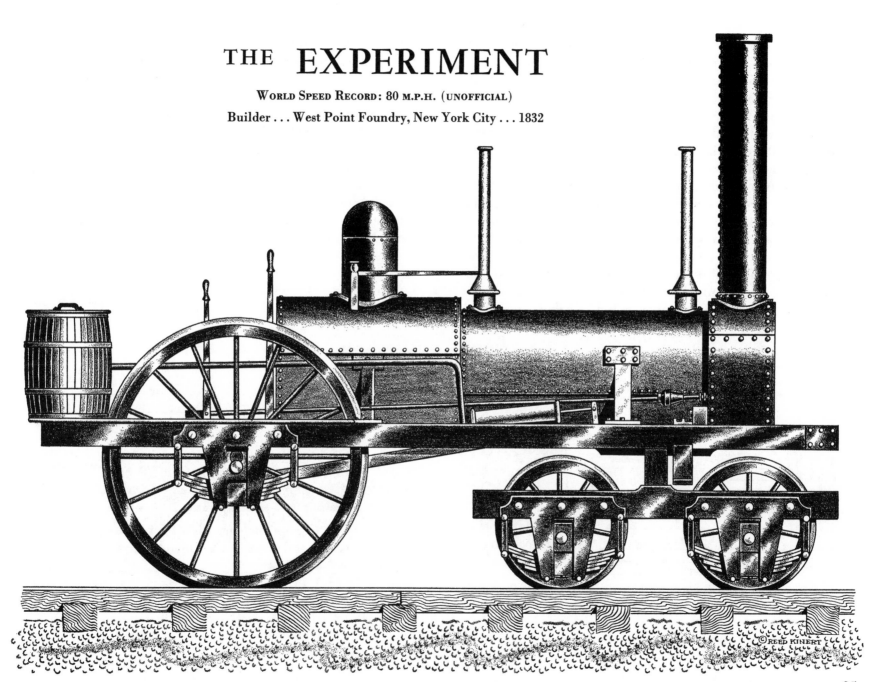

tinder and were pitched overboard. There was wild confusion as the clothing of one passenger after another caught fire.

Five miles from Albany the first stop was made and the resulting bumps and jerks caused hats to again be jolted from their heads and the passengers from their seats. Clark and some of the passengers raided a rail fence, cut and lashed even lengths of the rail to the coupling-chains to stop the jerking by holding the coaches rigidly apart.

This first trip of the *Clinton* had been well advertised and the whole countryside had turned out to see their first steam train. The track for nearly the entire distance from Albany to Schenectady was lined with farmers' rigs and many of the horses noses were close enough to the track to touch the monster when it came along. Puffing smoke and belching sparks, the noisy engine created untold confusion and havoc. There were many cases of smashed conveyances and shaken spectators,

while the little train steamed defiantly toward the end of its run, making the distance of fourteen miles in forty-six minutes.

The *DeWitt Clinton* managed to operate for a year before being retired in favor of an improved engine and parts of her were either sold or scrapped. The finding of one of its wheels around 1890, combined with an existing early scale drawing of the *Clinton*, together with a silhouette cut out of black paper with a pair of scissors, depicting the *Clinton* and two of her coaches at the scene of its first run, by a highly gifted man named William Brown, led to the construction of an exact replica which now stands in the Henry Ford Museum at Dearborn, Michigan.

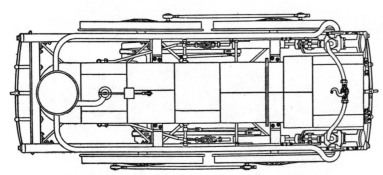

Top view drawing of the *West Point* engine—original concept

Side view drawing of the *West Point*

E. L. MILLER

GENERAL PURPOSE LOCOMOTIVE

Builder . . . Matthias Baldwin & Co. . . . 1834

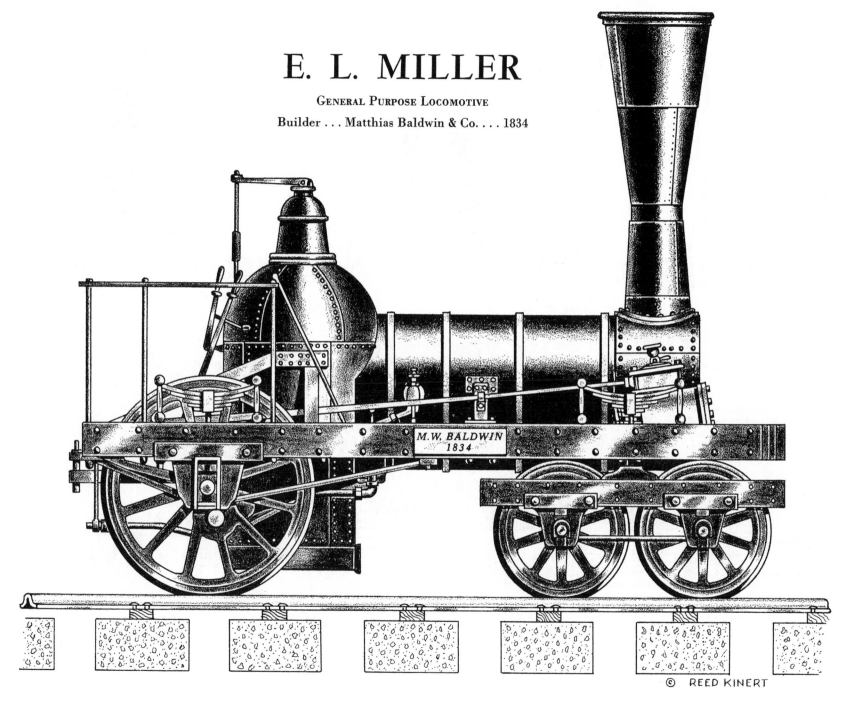

M.W. BALDWIN 1834

© REED KINERT

THE MONSTER

HEAVY FREIGHT TYPE 8-WHEEL ENGINE

Builder . . . Camden & Amboy Railroad Shops . . . 1836

CHAPTER IV
ANOTHER WATCHMAKER APPEARS

Now there appears on the scene another watchmaker, a Matthias W. Baldwin who had learned the jeweler trade in Philadelphia in 1819, then opened a shop of his own. Six years later he went into partnership with a David Mason, a machinist, to manufacture book binder's tools and cylinders for calico printing. Steam power for the shop became necessary so Baldwin designed and built a fine running engine which led the firm into steam engine building. Baldwin was asked if he could make a small locomotive, one powerful enough to pull a couple of small cars around a circular track at Peale Museum at Philadelphia. Baldwin thought he could. He set to work and on April 25, 1831, the little engine snorted and chuffed and pulled two cars, each seating four passengers, around the track at the Museum, and instantly it became the greatest attraction of the Museum.

All Philadelphia felt an urge to ride behind the tiny iron horse that made its own power as it moved; and among the many riders were men interested in the newly incorporated Philadelphia, Germantown & Norristown Railroad (now part of the Philadelphia and Reading system), six miles of strap rails of which had been built and on which horses were the motive power. These men asked Baldwin if he could build them a full size locomotive, an engine to haul freight and passengers on the new and growing railroad. Baldwin allowed it was possible. He visited the shops of the Camden & Amboy Railroad at Bordentown where the parts of the second English locomotive to be set on American rails, the *John Bull*, awaited assembly and Baldwin made an intimate and critical study of them, made notes and sketches then went home and so to work.

Although he had no patterns, no machine tools and there were but five mechanics available in the city of Philadelphia competent to do any work on a locomotive, Baldwin built his first full-size locomotive within six months and named it *Old Ironsides*—and it looked a lot like the *John Bull* engine. *Old Ironsides* was a four-wheel engine with the driving wheels in front of the fire box and the carrying wheels close behind the smoke box. Every locomotive built at this time was a noble experiment, either a new design, a collection of the better features of engines before it—or an outright copy.

On its first unofficial test run *Old Ironsides* must have been a disappointment to everyone. A Mr. Haskell who rode the engine with Baldwin related that the two of them alternately rode and pushed all the six miles to Germantown, and the trip took six hours! When they attempted a slight rise in the track on the homeward journey the engine stopped for want of steam so the two men availed themselves of a handy wood rail fence and soon had enough steam to make the grade and return to Philadelphia. One wonders how early railroading would have managed had it not been for the plentiful and handy wood rail fences along American right-of-ways! Used for patching all manner of breakdowns as well as an auxiliary fuel supply, there must have been many an irate farmer to contend with.

The railroad operators were frankly dismayed at the poor performance of *Old Ironsides* and were happy they still had their horses. Baldwin told them not to worry and he worked on the engine to such purpose that on November 24, 1832, the *Philadelphia Chronicle* could report that "The locomotive engine built by our townsman, M. W. Baldwin, has proved highly successful." The paper went on to say that the placing of the fire in the "furnace" and raising of steam took only twenty

1ST AMERICAN TYPE ENGINE

DESIGNER: HENRY R. CAMPBELL

Builder . . . James Brooks Locomotive Works . . . 1836

KINERT

The JOHN HANCOCK, one of the improved "Grasshoppers" built for the B & O in 1836, pulling its Imlay coaches toward the station for passenger loading. The upper deck awnings and seats on the coaches could be removed for bad weather trips.

minutes, then she moved with her tender from the station "in beautiful style, working with great ease and uniformity as she puffed to a point beyond Union Tavern then returned to the city, a total of six miles, without a halt." The paper added with pride that *Old Ironsides* had developed a speed of 28 miles an hour at one time.

The Philadelphia, Germantown and Norristown Railroad was immensely proud of its refurbished engine and immediately advertised that *Old Ironsides* would haul the cars every day only when the weather was good, the horses being held for rainy days and for snow. They didn't want their fine new engine to get wet! But this first Baldwin locomotive was better than they knew. Her maker continued to putter with her between her scheduled trips and a year later she astonished all by doing a mile in 58 seconds and 2¼ miles in three minutes and 22 seconds, either of which was a marvel in its time.

Old Ironsides weighed in at close to seven tons instead of the contracted five tons and the railroad directors would pay Mr. Baldwin only $3,500 for it instead of the $4,000 they had agreed to pay. This so disgusted Baldwin, after all the work he had put into the engine, that he vowed to never build another locomotive—but he never did anything else the rest of his life.

Other railroads were quick to hear of the success of *Old Ironsides* and sent their men to buy engines from Baldwin—and it seemed that each engine was better than the one before it. Baldwin perfected a steam-tight joint that would carry 120 pounds of steam and not leak, compared to the English joints made of canvas and lead that would not carry even 60 pounds of pressure for long. Many a Baldwin engine built in the 1830's and 40's was in use for the next three decades and more, *Old Ironsides* itself being in operation for more than twenty years.

While Baldwin was busy with his building of *Old Ironsides* the South Carolina Canal & Railroad Company down south had placed on the rails their third locomotive. Designed by Horatio Allen and built by the West Point Foundry in 1831, it was that foundry's fourth engine and was the first double-ended (articulated) locomotive in the world. Named the *South Carolina* and looking like two locomotives built together, back to back, it had each pair of axles mounted on a swivel, which also carried a single cylinder connected to a driver axle crank. The *South Carolina*, while a noble experiment, spent most of its time in the repair shop and proved to be an impractical freak, ponderous and heavy, so the *Best Friend* locomotive parts were salvaged and rebuilt with straight axles, outside cylinders and cast iron wheels. Much changed in appearance and renamed the *Phoenix*, she returned to service in October 1832, did a creditable job for several years.

When the South Carolina Canal and Railroad Co. was ready to order their fourth locomotive in 1833, Mr. E. L. Miller was entrusted to journey to Philadelphia and attempt to talk Mr. Baldwin into building another engine. The two men examined several locomotives that had been imported from England and it was concluded that Baldwin could build a better engine. Named the *E. L. Miller*, the new locomotive had outside wooden frames in the English manner and a single pair of drivers behind the boiler and a four wheel truck under the smoke box. The cylinders were set on top of the frames at the sides of the smoke box and transmitted the power to half cranks inside the driving wheels. This engine was delivered in March 1834 and became one of the most famous locomotives of its day. Its form was standard with Baldwin until the need for heavier power decreed radical design changes.

The *E. L. Miller* had very little that was decidedly original but old forms were combined into a shape that produced the best engine then built. The most conspicuous feature of the engine was simplicity of parts. A boiler that anyone could understand and that any boilermaker could repair, a pair of cylinders securely fastened between the boiler and frame, a valve motion having no mystery, a running gear that combined strength with simplicity.

THE TOM THUMB
AMERICA'S 1st STEAM LOCOMOTIVE

WHEELS, DIAMETER 30 1/4"   WEIGHT TOTAL ENGINE 10,800 LBS.

Builder ·· Peter Cooper ·· 1830

REED·KINERT

BEST FRIEND of CHARLESTON

America's First Regularly Scheduled Passenger Locomotive

Builder . . West Point Foundry . . 1830

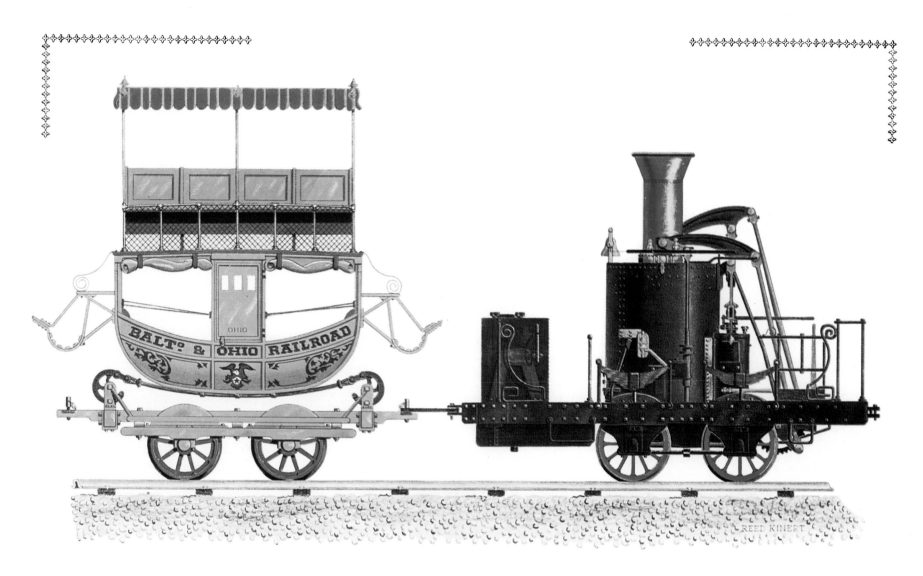

ATLANTIC

FIRST GRASSHOPPER TYPE LOCOMOTIVE

Builder . . Davis & Gartner . . 1832

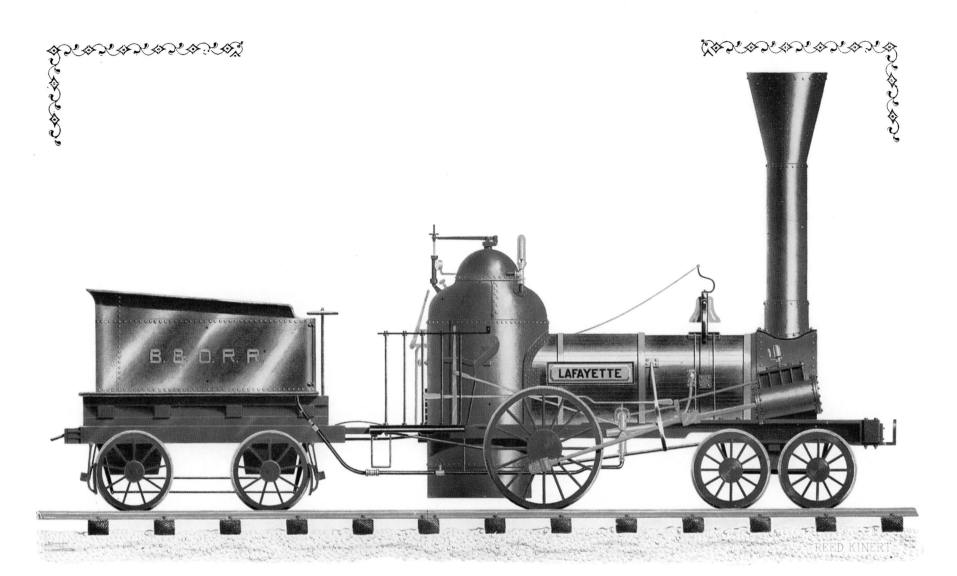

THE LAFAYETTE

DRIVERS, DIAMETER 48"   WEIGHT, TOTAL ENGINE 28,300 LBS.

1st HORIZONTAL-BOILERED ENGINE ON B&O R.R.

Builder ·· William Norris & Cᵒ ·· 1837

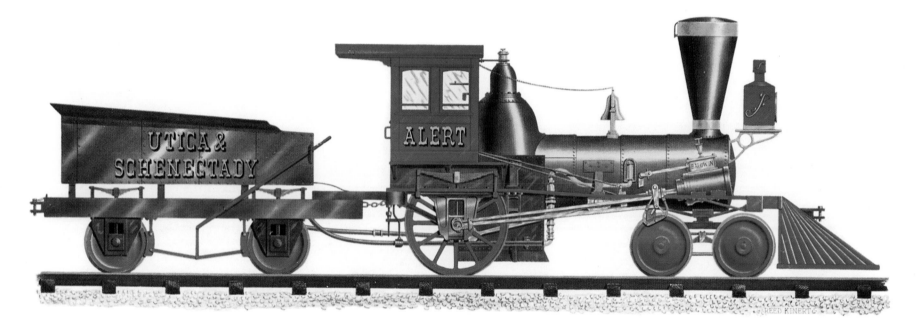

ALERT
LIGHT PASSENGER ENGINE

Builder . . Matthias W. Baldwin & Co. . . 1837

WASHINGTON
SIX-COUPLED FREIGHT ENGINE

Builder . . Matthias W. Baldwin & Co . . . 1847

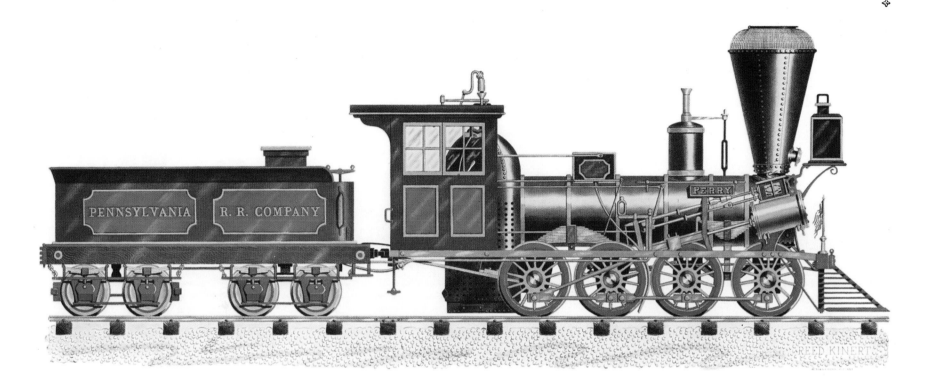

THE PERRY

A FREIGHT ENGINE OF GREAT POWER

Builder . . Matthias W. Baldwin & Co. . . 1848

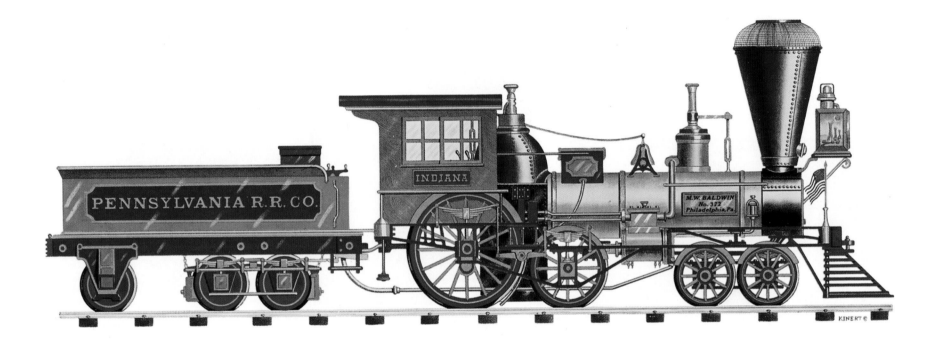

INDIANA

SINGLE-DRIVER FAST PASSENGER ENGINE

Builder . . Matthias W. Baldwin & Co. . . 1849

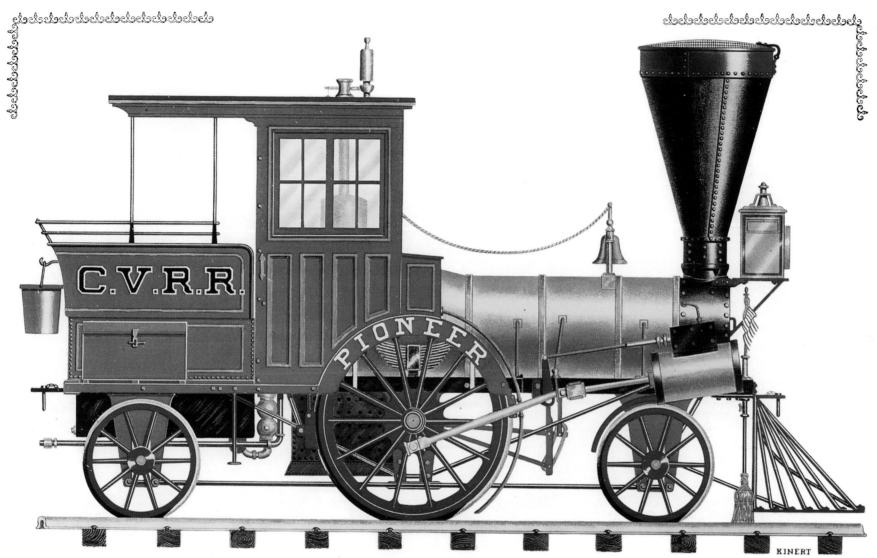

PIONEER
LIGHT PASSENGER LOCOMOTIVE

Builder . . Seth Wilmarth of Boston, Mass. . . 1851

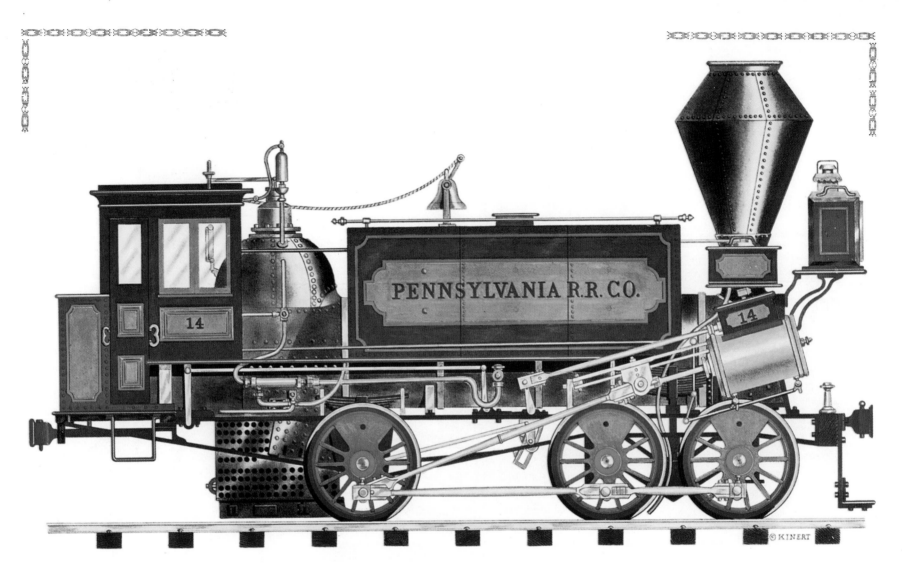

TANK TYPE
SIX-COUPLED SWITCHING ENGINE

Builder . . Matthias W. Baldwin & Co. . . 1852

THE TIGER
FAST PASSENGER ENGINE

Builder . . Matthias W. Baldwin & Co. . . 1856

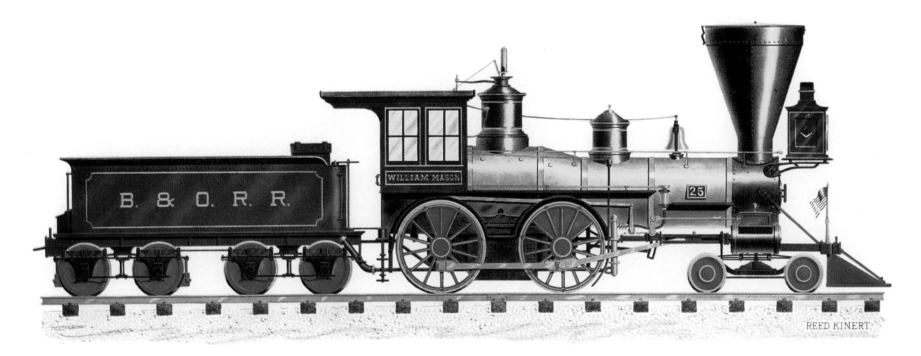

WILLIAM MASON
FAST PASSENGER ENGINE

Builder . . Wm. Mason Machine Works . . 1856

C.P. HUNTINGTON
LIGHT PASSENGER ENGINE

Builder . . Danforth Cooke & Co. . . 1863

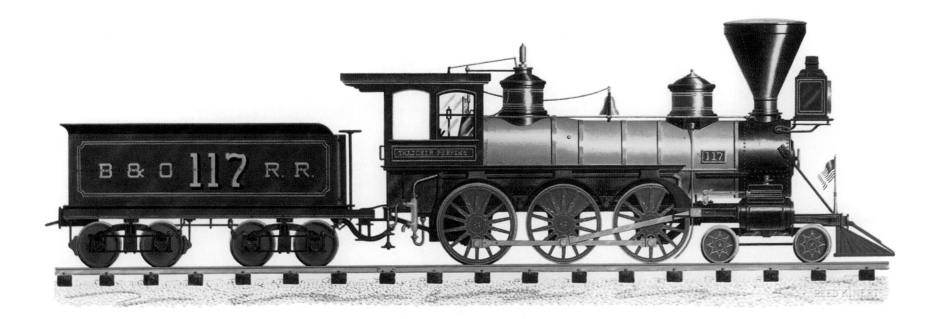

THATCHER PERKINS

Mountain-Type Passenger Engine

Builder . . Baltimore & Ohio Shops . . 1863

FAST EXPRESS LOCOMOTIVE

Built at

ILLINOIS CENTRAL RAILROAD SHOPS, 1866

Samuel J. Hayes, Supt. of Machinery

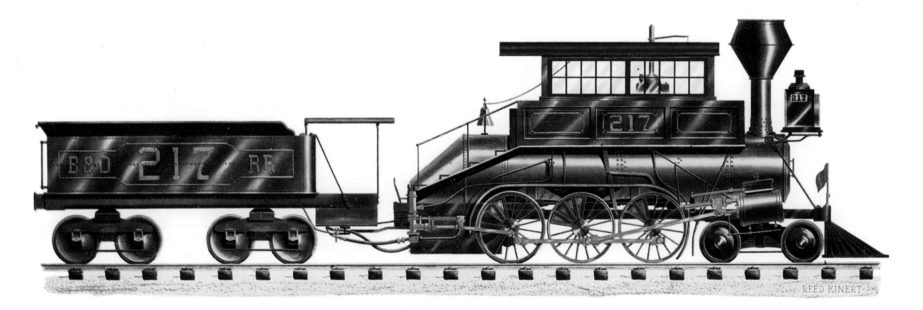

TEN-WHEEL CAMEL
POWERFUL MOUNTAIN-TYPE FREIGHT ENGINE

Builder . . Baltimore & Ohio Shops . . 1873

CHAPTER V

DURING WHICH STEADY PROGRESS IS MADE

Although the *DeWitt Clinton*, America's fifth locomotive, was no world-beater from any standpoint, her designer John Jervis was not to be denied. It was he who, as Chief Engineer of the Delaware & Hudson, had ordered the first four locomotives to be brought over from England. While the engines were not a success, Mr. Jervis, quite a mechanical genius in his own right, was quick to see their faults, but became on the spot, a convert to steam.

The very next locomotive he designed after the *Clinton* was the engine *Experiment*, sometimes called *Brother Jonathon*. Being dissatisfied with the rigidity, poor turning and inherent destructable characteristics of the ponderous four-wheel English engines, Jervis designed a machine which consisted of a four-wheel lead truck and only two drive wheels, the four forward wheels being connected to the frame on a pivot. Some reports state that Jervis rebuilt one of the four imported engines mentioned in Chapter I to build his *Experiment*, removing the rigid front axle, with its single pair of wheels and substituting the front truck with two axles and 4 wheels. Some of the parts might well have been cribbed from the English engines but the result was better distribution of the engine's weight and also shortened the wheelbase, permitting the machine to take the curves in the track with less wear on wheels and rails. Cabless, primitive looking and scorned by English and American locomotive builders of the day the Jervis designed, West Point Foundry built (their fifth and last engine) *Experiment* proved to be the world's first mile-a-minute locomotive. It was by far the most advanced of six different engines built in the U.S. in the year 1832, and was far better than anything England had built.

David Matthew who, while working for the West Point Foundry, had supervised construction of and then been engineer on the *DeWitt Clinton,* was also the first to pilot the *Experiment*. He wrote "With this engine I have crossed the Mohawk & Hudson, fourteen miles in thirteen minutes (65 mph) making one stop for water. I have tried her speed upon the level straight line and have made one mile in forty-five seconds by the watch. (A remarkable 80 mph!) She was the fastest and steadiest engine I have ever run or seen run and she worked with the greatest of ease."

After these marvelous test runs the locomotive was more copied than criticised and one can see the *Experiment* locomotive of 1832 in many later engines, including Baldwin's second locomotive, the *E. L. Miller* of 1834. This type of engine became practically standard with the Baldwin Works for several years. When one reflects that the 80 mph of the *Experiment* was accomplished sixty-six years before the automobile was even invented and seventy-one years before the first airplane flew, the feat is all the more impressive.

Despite the amazing speed of the *Experiment* its four-wheeled truck or bogie as it was named (now called pilot wheels) was the most salient feature of the engine. It was this bogie that guided or piloted the engine down the rails and caused it to hold the rails so well. The bogie proved to be one of the greatest of early inventions for without it locomotives could not have grown in size and yet be flexible over rough or curving track.

While John Jervis was busily building and testing his *Experiment* in New York our Mr. Phineas Davis was busy down in Baltimore town. The Baltimore & Ohio Railroad had been fortunate, not only to obtain the services of Davis who had set up a locomotive shop in Baltimore, but those of his versatile assistant engineer, Ross Winans. Winans' first contribution to railroading was to place the wheel flanges of cars and locomotives on the inside of the track rather than the outside for better rail-holding. Winans too saw the advantage of having the axle turn with the wheels, a feature still in use on railroads today.

The B & O rails had reached Frederick, Maryland, sixty-one miles west of Baltimore, on December 2, 1831 and was building towards Harpers Ferry, Virginia. The locomotive *York* was found lacking in climbing the grades of the more mountainous western reaches of the railroad so Phineas Davis and his partner Gartner, and Ross Winans of course, kept steadily at work trying to improve on steam locomotion.

In the summer of 1832 this trio of mechanical brains placed on the rails of the B & O a second steam engine, the *Atlantic*, which weighed about 14,000 lbs., then considered a little beyond the safe weight for running on strap rails. In this developmen and construction of this larger and more powerful engine Peter Cooper acted as consultant and was of great help. The little *Atlantic* showed herself from the first to be most capable and was soon placed on regular passenger service to Parr's pulling a brigade of five loaded passenger cars of the type Ridge, forty miles west of Baltimore and return each day, pulling a brigade of five loaded passenger cars of the type shown in the color illustration of the *Atlantic*. Although these cars were exact copies of stage coaches, complete to leather springs, and built by coach-maker Robert Imlay of Philadeledphia, they more than a little resembled facied-up gravy boats! The awning tops were detachable, being braced-up with threaded pipe, while the seats atop the coach could be unscrewed, for the top deck was used only in the finest of weather.

Although she looked like a homely little toy, the *Atlantic*
could make the daily round trip of eighty miles on one ton of coal and was the most efficient fuel consumer that America had yet seen. In 1832 Jonathan Knight, the chief engineer of the B & O, reported that the *Atlantic* hauled a gross load of 50 long tons from Baltimore to Parr's Ridge up an average grade of 37 feet to the mile, at a speed from 12 to 15 mph. Daily expense of the 80-mile round trip was $16, which included one ton of anthracite coal at $8.00 a ton; engineer $2, fireman $1.50, oil and packing 50 cents, estimated wear and tear and interest on cost $3.00, water station expense $1.00. The engine did the work of 42 horses whose daily expense was $33.00. Initial cost of the engine was about $4,500.

Quite formidable in appearance, the *Atlantic* was the first of a series of "grasshopper" engines, a type so called because of the pair of long perpendicular driving rods, resembling the hind legs of a grasshopper, that connected her two cylinder rods with the two drive wheels through the use of gears.

While the Davis built *York* had been the first locomotive to use steel springs, they were also used on the *Atlantic* for it was found they "greatly diminished the jar and consequent injure to the road, as well as the engine." Observant Ross Winans soon tried steel springs on passenger and burden (freight) cars and found that the capacity of the cars was increased, while they were easier to haul.

The original *Atlantic* was in service for only about four years and disappeared, her parts probably being used in other engines.

While many early American built locomotives could be called at least partial copies of English design, this certainly did not hold true of engines turned out by Phineas Davis. Influenced by the *Tom Thumb* to build his engines with upright boilers, his products, like Peter Cooper's *Tom Thumb*, were as American as hominy and apple cider.

The *Atlantic* was one of the five successful locomotives known to have been built in the U.S. in the year 1832, including

THE HERCULES

FIRST LOCOMOTIVE WITH DRIVE-WHEEL EQUALIZER

Built by Garrett & Eastwick, Philadelphia . . . 1837

the *Phoenix*, rebuilt from the damaged *Best Friend of Charleston*. The others: the *South Carolina, Old Ironsides* and the *Experiment*, all illustrated in this book. After 1832 such a rapid pace in locomotive building was set in the United States that individual mention becomes impossible and the exact number built is not known so we will, of necessity, choose some of the more outstanding engines for comment in the following chapters.

Railroads began multiplying and expanding. It seemed as if every state, every city, every one-horse village in the eastern United States was bound it should be on a railroad, even if it had to build one.

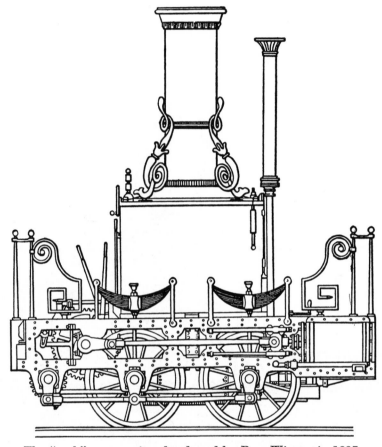

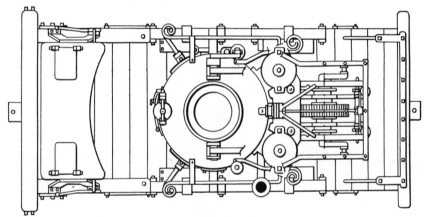

Top view of the *Atlantic* "grasshopper" type locomotive.

The "crab" type engine developed by Ross Winans in 1837.

United States—1830 to 1840

	Population	Miles of Railway
1830	12,866,020	23
1831	13,252,000	95
1832	13,571,000	229
1833	13,924,000	380
1834	13,319,000	633
1835	14,743,000	1,098
1836	15,127,000	1,273
1837	15,532,000	1,497
1838	16,037,000	1,913
1839	16,540,000	2,302
1840	17,069,453	2,818

CHAPTER VI

THE "GRASSHOPPERS" HAVE THEIR DAY

History records show that at least sixteen vertical-boilered "grasshopper" type engines were built for the Baltimore & Ohio Railroad: the *Atlantic* of 1832, followed by the *Traveller, Arabian, Mercury* and *George Washington*, all built by Davis and Gartner in 1834 at the Mount Clare shops in Baltimore. Two unnamed grasshoppers were also built by a Charles Reeder of Baltimore, probably under sub-contract and using Davis drawings for the railroad was converting to steam entirely, and as quickly as possible. One of these two unnamed engines exploded before delivery in November of 1834, killing Mr. Reeder. Then, in 1835, four more "hoppers," the *James Monroe, John Quicy Adams, Thomas Jefferson,* and the *James Madison,* were listed, built by George Gillingham and Ross Winans. Five others, the last of this class engine, the *Andrew Jackson, John Hancock, Phineas Davis, Martin Van Buren* and the *George Clinton* were built in 1836, also by Gillingham and Winan.

The principle difference between the *Atlantic,* first of her class, and the later grasshoppers was that power was transmitted to only two wheels on the Atlantic while the later engines had their power delivered to all four wheels by a main drive and connecting rods, but all these engines had one thing in common. With their vertical cylinders and their walking beams which resembled giant rocker-arms, they all looked like huge grasshoppers walking down the rails when under way.

The grasshoppers caused a great storm of controversy among locomotive builders and designers. Phineas Davis had been killed in August of 1835 when a loose rail sent his grasshopper engine into a ditch while traveling at more than twenty-five miles an hour on the newly opened Baltimore-Washington, D.C., branch and his partner, Mr. Gartner, quit building locomotives in grief. Mr. Davis was a man who was either loved or respected, or both, by all who knew him. George Gillingham, Superintendent of Machinery for the road, and the indefatigable Ross Winans took over construction of B & O locomotives. Gillingham and Winans steadily argued in favor of the grasshoppers, claiming that the upright boiler preserved the water at a better level on steep grades and that the horizontal-boilered types frequently burst their tubes, which was not only expensive but stopped the engine by drenching its fire and so caused traffic upon the railroad to stop.

The grasshoppers were unique to the Baltimore & Ohio Railroad as the road eagerly used every "hopper" built in their race to replace the more costly and slower horses on their line, until they found a locomotive on the open market that could out-perform the homely little teakettles. The engines were fantastically economical to operate and burned anthracite coal which gave a much hotter fire than did wood, and the engine grates were easy to replace. One little grasshopper was credited with pulling the Frederick passenger train for fifty successive days, eighty-two miles a day, without a lay-off or repairs of any kind. Her daily expenses for the 42-mile round trip came to $13.25; $1\frac{1}{4}$ tons of coal $7.50, oil and packing 50 cents, engineer's wage $2, the fireman's $1.50, an interest charge on her cost of 75 cents, and a contingency fund, $1.00. It was estimated that she did the work of 113 horses, and at a vastly less cost.

The little grasshoppers did very well in load pulling for

NAZEPPA

"CRAB" TYPE LOCOMOTIVE

Builder . . . Ross Winans & Co. . . . 1837

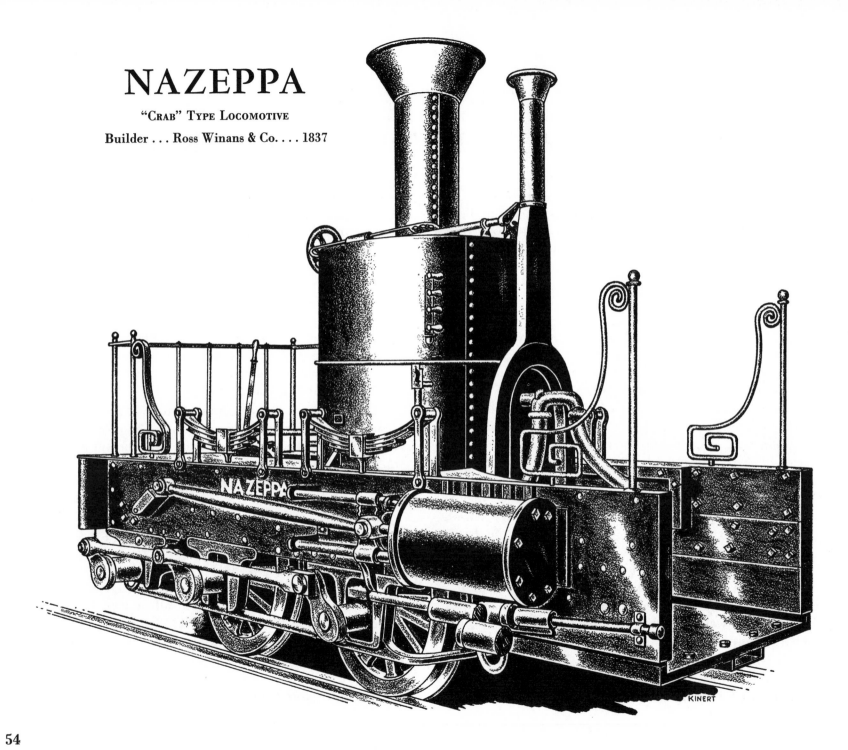

GOWAN & MARX

WORLD RECORD FREIGHT HAULING ENGINE: 423 TONS

Builder . . . Eastwick & Harrison, Philadelphia . . . 1839

their time, too. The distance between Baltimore and Washington, D.C., was forty miles and a grasshopper, drawing three Ross Winans designed eight-wheeled cars of thirty-four foot length, containing 140 passengers together with their baggage and a baggage car could travel at sixteen to twenty miles an hour. With 75 to 100 passengers the trip could be made in one hour and a half, with speeds up to 40 mph. Although the engines were capable of maintaining a schedule it was almost impossible to do so during that time. There were no stations between towns and the towns were far apart so if a passenger wanted to board a train they merely flagged the train down anywhere along the line. Sometimes the little grasshoppers made more stops than a city bus does today.

Although not much bigger than a handcar the grasshoppers could scat down the rails if called upon to do so. In one instance, in 1836 it was proposed that a speed run be made between Washington and Baltimore to deliver an important dispatch of mail. The engineer claimed he could make the journey in forty minutes, an average speed of sixty miles an hour. Bets were laid with wild abandon and the run was begun to a great cheering of the assembled crowd, but after the engineer had started and was puffing his steed merrily along at better than sixty miles an hour, the fireman became alarmed at the high speed and quickly turned the cocks which allowed steam to flood the fire-box and put the fire out!

The grasshoppers had their day but were doomed. Their vertical cylinders and long walking beams imparted a bobbing motion and the vertical boiler made the engine top heavy, a defect which was aggravated by the short wheel base. They had reached their practical limit in size and the railroads were constantly requiring larger engines to haul ever increasing traffic. Also in addition to larger boiler size, both actual and potential, the horizontal-boilered engines had a lower center of gravity than the grasshoppers, no little aid in holding the track better, especially on curves.

Of the many grasshoppers constructed, three have survived. The earliest, the *John Quincy Adams*, built in July, 1835, is now exhibited in Carillon Park at Dayton, Ohio. Restored and repainted, it is the oldest complete American built locomotive in existence. The remaining two, the *Andrew Jackson* and the *John Hancock*, built in 1836, are housed in the B & O Museum at Baltimore. All three engines were in active use at the Mount Clare station in Baltimore as recently as 1892, serving as switching engines! In 1893 the *Andrew Jackson* was altered to resemble the first grasshopper built, the *Atlantic*, and remains so altered today.

It is of interest that at least one other grasshopper locomotive was built by Gillingham and Winans. Named the *Columbus*, this little known example was made in 1836 for the Leipzig to Dresden Railroad in Germany and quite probably was the first American locomotive built for export.

In an effort to stave off the inroads made by horizontal-boilered locomotives, Ross Winans, strong exponent of the vertical boilered engine, developed, in 1837, for the B & O, a locomotive type that came to be known as a "Crab." While the boiler was upright, the cylinders were placed horizontally and it was their drive action which gave the appearance that the engine was running backwards, hence the name "crab." Placing the cylinders horizontally helped to overcome some of the defects of the engine, but they were unable to attain the high speed of other types of locomotives built during the first decade of the railroad era. On the B & O Railroad the capacity for hauling freight was more important than speed and Winans improvements on the engines were made mostly to increase the tractive power.

* * * *

CHAPTER VII

THE WONDROUS STEAM ENGINE GROWS

By 1837 Philadelphia already had become a locomotive building center of importance. Three firms—William Norris, Eastwick & Harris and Mathias Baldwin were building up reputations as builders of reliable engines. At the same time, over in Paterson, New Jersey, the locomotive firm of Rogers, Ketcham & Grosvenor were building engines of note. This firm eventually became the Rogers Company. In that day Norris was perhaps the brightest star of the group. He built his engines in the traditional American manner, with the frame inside the wheels while Baldwin and the others mostly clung to the English style of outside frames. In 1836 Norris had built an engine, a six-wheel type, composed of a four-wheel bogy or pilot and two rather large drive-wheels. In appearance it looked a lot like the Jervis engine *Experiment* except for the frame being inside the wheels. This new engine, named the *George Washington*, ran so well on its trial trip that Norris, who was always regarded as a reckless dare-devil person, decided to test his engine all-out and took it to the bottom of the steep inclined plane of the young Columbia Railroad, just west of the Schuylkill River.

These inclined planes were peculiar to early eastern railroads in mountainous country and worked in this manner. Where the mountains were too steep or it seemed impractical to go around a mountain an incline would be built right up the mountainside. The train or brigade of horse or locomotive drawn cars would travel along the relatively level track until coming to the mountain with its incline whereupon the cars would be pulled up the steep incline, either by a stationary steam engine or by horses through a system of pulleys and a cable or huge rope, and then, at the top of the incline, the cars would be hitched to other horses or another locomotive and be transported to the next inclined plane and so on until the method was reversed and the cars let down the other side of the mountain in the same manner. Sometimes inclined planes were used in conjunction with canals and even a combination of railroads, inclined planes and canals were used.

The Schuylkill inclined plane that Norris chose to test his engine was quite steep. The length of the incline was 2800 feet, the grade 369 feet to the mile or a 1-foot rise in 14.3 feet. Several hundred people turned out for the gala occasion and when all was ready Mr. Norris started the engine at the very bottom of the incline, but after proceeding a few feet the wheels slipped and the engine slid back to the starting point. Examination of the rails disclosed that eager workmen, with bets against the locomotive, had heavily oiled the track. Norris quickly sprinkled sand on the rails and proceeded to steadily climb the engine which gained speed as she advanced to the very top of the incline. To his unbounded delight the *George Washington* had climbed the incline in 2 minutes and 24 seconds while pulling a tender with coal and water and two passenger cars loaded with 53 passengers, a total weight of 31,270 pounds!

Performance of the *Washington* was so astounding that Norris' reports of it were met with disbelief from all points. Even after the feat had been repeated in the presence of many more witnesses it could scarcely be credited. Eight months after Norris had demonstrated that a locomotive could not only climb an ascending grade by its own power, but could also haul a train up, a Mr. A. G. Steere, of the Erie Railway, sent a long

FAST EXPRESS ENGINE

MOHAWK & HUDSON RAILROAD

Designed and Built by David Matthews . . . 1840

FREIGHT LOCOMOTIVE ENGINE

SIX WHEELS — CONNECTED TYPE

Builder . . . Matthias W. Baldwin & Company . . . 1842

communication to the *Railroad Journal* of March 11, 1837, proving by elaborate algebraic formulae that the *Washington* did not climb the hill because it could not; and that no other locomotive ever could climb an ascending grade by its own power, and fully exposed "the deeds done in flagrant and open violation of the laws of gravitation." It was rather difficult, however, to maintain that a thing could not be done when it was being done daily and the controversy finally died out.

An immediate result of this astounding performance—at first received with disbelief, but later proven in full reports, was that Mr. Norris was swamped with orders for engines. The directors of the Baltimore & Ohio alone ordered eight locomotives of Norris. Delivery of these began with arrival of the *Lafayette*, which was placed in service April 1837, making it the first horizontal-boilered locomotive on the B & O. The *Lafayette* was a 4-2-0 type engine—a front pilot truck or "bogie" with two axles and four wheels, followed by two large drive wheels and was almost a twin to the earlier *George Washington*. This wheel arrangement resulted in better weight distribution and also shortened the engine's wheelbase, permitting the machine to take the curves with less wear on the wheels and rails.

Unlike the last Baltimore & Ohio "grasshopper" models, the *Lafayette* did not wear a cab for her crew. Passengers complained that cabs were likely to cause locomotive operators to be less vigilant, while the truth probably was that the passengers couldn't see the water gauge with a cab in the way and they had soon learned that a low water supply meant an explosion. The engineers themselves were mostly set against cabs too, for they wanted plenty of jumping room in case of an impending accident. You will notice in the color illustration that Mr. Norris had taken the safety precaution to band hardwood horizontal strips, or staves, onto the boiler. This was to prevent chunks of metal from flying too freely and far in case the boiler exploded. This safety device came to America on the first English locomotives to be shipped over here, indicating that the English, too, had had trouble with explosions, and the wooden strips were quite a common sight on both American and English locomotives of this period.

It was about this time in the history of steam locomotive development that designers turned their attention to the size of driving wheels. It was quite apparent to them that the larger the driving wheels, the farther the locomotive would move with each revolution of the drive wheels. Engines with huge driving wheels as large as eight feet in diameter were tried, but it was soon found that too much power was required to turn these outsize wheels and added horsepower meant tremendously heavy-weight locomotives, quick to chew up flimsy rails and destroy road beds. A basic design rule was evolved that the diameter of driving wheels in general was to be governed by the speed for which the engine was to be designed. Thus a 30 mph locomotive design usually wore drive wheels of approximately 30 inches, while a 50 mph locomotive used wheels of about 50-inch diameter.

William Norris built several engines of the *Lafayette* type for other railroads in addition to the eight for the B & O. They became known as the class "B" engine, as designated by Norris, and were built both with and without cabs. These engines were so superior to all previous American locomotives that the type was copied unabashedly by almost all the other engine builders. Baldwin even lightened the outside frame of his locomotives in an effort to make them more flexible but he still left the frame outside the wheels.

Soon surplanted on the main line by still larger and more powerful engines, the *Lafayette* and her sister engines were relegated to switching duties and all of them eventually disappeared, either scrapped or destroyed in the Civil War as the railroad changed hands several times between the North and South.

*   *   *   *

CHAPTER VIII

CANALS, ROADS AND RAILROADS

By 1811 there were 37,000 miles of post roads suitable for wagon and stage coach use in the United States, including the great National turnpike, begun in 1806 and built from Cumberland, Maryland, to Stuebenville, Ohio. States followed the example of the National government by appropriating money for highways, while turnpike and plank road corporations were privately financed. In time it became possible to travel from Boston to New York by stage coach, a distance of 270 miles, in twenty-five and a half hours; and from Baltimore to Louisville by stage and steamboat in six days and eight hours, including a total of twenty hours detention at various places.

But a freight tariff based on a rate of a hundred dollars a ton between Albany and Buffalo, as an example of shipping rates of the time, was not calculated to foster trade. When the necessities of traffic became so urgent that they could no longer be disregarded the people turned to the building of canals as a source of relief. In 1762 some citizens of Pennsylvania applied to the legislature for a charter for a Schuylkill and Susquehanna Canal to connect the rivers of those names. This canal, four miles long, finished and operated in 1794, was the first projected and the first chartered, while the South Hadley & Montague Canal, five miles long, around the rapids of the Connecticut River at South Hadley, Massachusetts, opened in 1793, was the first operated in the United States.

The sentiment in favor of canals thus started spread gradually until the astonishing results accomplished by the Erie Canal, completed in 1825 and connecting Albany on the Hudson River, with Buffalo on Lake Erie, fanned it into a furor. Immediately all sorts of possible and impossible canal projects

were advanced, with the result that by January 1, 1835, forty-eight canals totaling 2,617 miles were in use. At the climax of the canal period five thousand miles of these artificial waterways were in operation. Public opinion, which had so reluctantly taken this advance step, now assumed that the last word had been spoken on the subject of transportation; that the canal was the apotheosis of engineering skill.

One can easily imagine the furor caused when the first railroads were proposed. Here was an upstart that would upset the apple cart by breaking directly into the middle of the established order of things. From the very first our railroads were strongly and bitterly opposed by freight wagon owners, stage coach, canal boat and canal owners, as well as inn keepers and toll road owners and they were condemned by the irate citizenry if they tried to pass through their towns.

With the advent of steam locomotives to replace the horse-drawn cars, the furor against railroads increased mightily. In 1836 the people of Newington, Connecticut, drew up a protest against a railroad that had been surveyed through the town, declaring that they were a peaceable, orderly people and that they did not want their quiet disturbed by steam cars and the influx of strangers. Dorchester, Mass., in a town meeting assembled in 1842 instructed its representatives in the legislature to "use their utmost endeavors to prevent, if possible, so great a calamity to our town as must be the location of any railroad through it."

Thus, to their later dismay, many protesting towns found themselves by-passed as gleaming rails ever reached out over longer and longer distances to faraway places beyond the

61

horizon. Leaving the wagons to wallow in the mud and the canal folk to splash in their water-filled "ditches," the railroads made a history of mechanical progress that was unique to the world.

The restrictions placed on the Utica & Schenectady Railroad by the New York State legislature, induced no little by the highway and canal interests, was a typical case of unfair practice against early railroads. The railroad, seventy-eight miles in length ,was opened on August 1, 1836, with two horse drawn trains of three coaches each. Since the rail line was in competition with the Erie Canal, which it paralleled, the State decreed that the railroad could charge no more than four cents a mile for passengers and that no freight except baggage could be carried in the summer months! Also the line was forced, by law, to buy out any turnpike owners who wished to sell. This sort of treatment towards the railroads was typical instead of an exceptional case.

In an effort to cut costs by operating more swiftly and cheaply, the Utica & Schenectady Railroad placed an order with Mathias Baldwin for six steam locomotives, for even with restrictions of all kinds, their traffic had increased steadily. First of the six delivered in 1837 and set on the rails was the doughty little 4-2-0 *Alert*, which shows an early beginning at bridging the gap between "teakettle" engines and a really practical type locomotive.

The Erie, as well as all other northern canals, froze over solid during the winter months. The Utica & Scheneactdy, along with other northeastern railroads with restrictions against them, looked forward to below freezing weather, hoping to haul enough additional freight and passengers during that time to make up deficits during that frozen-in monopoly period. While ice could not freeze the locomotives and steam was piped into the tender to keep the water from freezing, snow-bound trains were commonplace. In the cars, at first unheated, travel was no lark and winter travel by train was a rugged adventure.

The *Alert* was delivered without a cab for the crew but the intense northeast weather soon compelled the railroad to install one, so she became one of the first Baldwin engines to be so fitted.

The *Alert* was also one of the first Baldwin engines to sport a headlight although it was not newly delivered with one. Early headlights first used candles, one or two, and then were converted to use whale-oil, for petroleum with its derivative coal-oil had not been discovered. Later and until the advent of electricity and generators, locomotives were fitted with very bright but not always reliable acetylene gas headlights.

The *Alert*, like all engines of her day, was lubricated with tallow-lard. Atop the firebox in the cab of every engine sat a warmed tallow-pot. Since it was the fireman's duty to keep the tallow hot so the engineer could lubricate the engine along the line, he was soon nicknamed "tallow-pot." If you look atop the steam chest of cylinders on early locomotives you will find the hand-operated fitting for "lubing" the moving parts therein.

Along with the fireman, the engineer was also in railroad fashion, soon nicknamed. Since the engineer was the "head" or boss of the engine and many of them used, or "hogged" more steam than they needed at times, they earned the name "hoghead"—and the names for both enginemen still stick.

The *Alert*, Baldwin's thirty-seventh locomotive, was typical of Baldwin built engines for it was made with an outside frame but, like all Baldwin engines was very well made and, before his death in 1866, his works had built more than 1,500 engines and was well on the way to the top position which his firm attained and held for many years after. The *Alert* served the Utica & Schenectady R.R. (now part of the New York Central) faithfully for several years and then, to make way for larger engines, was sold to the Michigan Central. The Central later sold her to the Galenta & Chicago Union Railroad. Now third-handed but still chipper and wearing the name *Pioneer*, the ten-ton Baldwin-built locomotive was already eleven years old

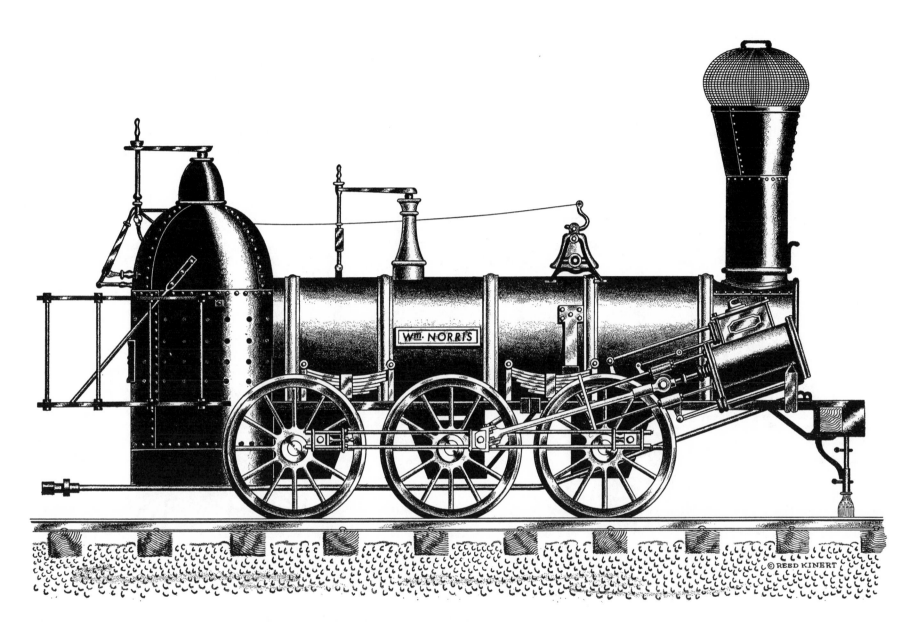

SIX-COUPLED FREIGHT ENGINE

Builder . . . Wm. Norris, Philadelphia . . . 1842

when sailors and dock workers struggled to drag the iron horse ashore from the brig "Buffalo."

On October 25, 1848, the *Pioneer*, with a small group of town dignitaries and the lines' directors seated on a flat-bed freight car behind it, chugged five miles west of town and back to introduce steam railroading to the little city of Chicago which grew to be America's greatest rail center. The trip was historic, too, for the fact that it marked the beginning of what is now the far-flung Chicago & North Western Railway. Since 1934 the *Alert* or *Pioneer* of you please, has been in dignified retirement at Chicago's Museum of Science & Industry but can still roll under her own steam when called upon.

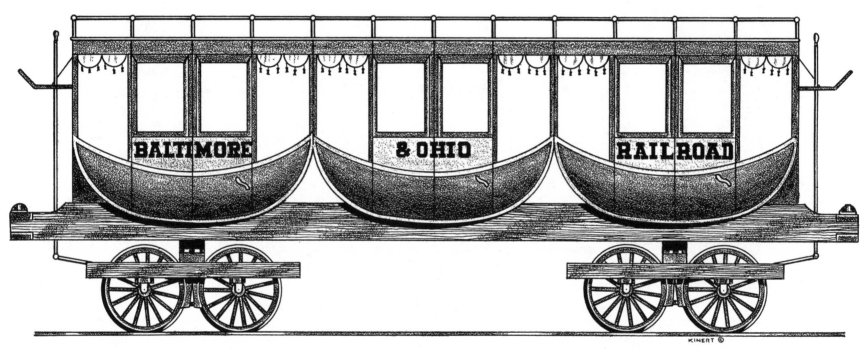

First double truck 8-wheel passenger car. Built by Ross Winans in 1831.

CHAPTER IX

BELLS, WHISTLES, HEADLIGHTS AND OTHER "EXTRAS"

Until about 1850 every locomotive built was mostly a unique one-of-a-kind type. No matter how good a newly designed engine was it seemed to be soon antedated by a more superior engine. Gradually, however, around 1850, locomotives began to assume a somewhat standard appearance. At least one could look at an engine and tell what its specific job was and one could, in many cases, tell at a glance who the builder was. Bells, whistles, headlights and sand boxes (domes) became standard fixtures perched atop the boiler and cowcatchers, later called the pilot, were affixed to all but switch engines. Drive wheels came to be counter-balanced with lead weights to offset the drive-rod weight which had been pounding the track, and the locomotives themselves, to pieces.

The most distinctive features of American steam locomotives—bell, pilot and headlight—are due largely to the cheap, hurried construction of the early railroads, and the great open and wide American space. Early American railroads were unfenced, since few of them had the money or inclination to build mile after mile of fence along both sides of the track, and streets and highways were usually crossed at grade. Livestock, as well as people, walked along the track at will—and the railroads, not, as in Europe, the owner, were liable if a trespassing animal was injured or killed. Collisions between cattle and locomotives was not only expensive—as the value of livestock seemed to increase sharply when it tangled with a train—but they were likely to derail the locomotive. The first locomotive with a pilot was the Camden & Amboy's *John Bull,* an English locomotive built by Stephenson and shipped over to this country in 1831. When it was rebuilt in 1833, Isaac Dripps, the C & A's ingenious master mechanic, added a low,

two-wheel pilot truck with long pointed rods jutting out in front of it to spear any animal it hit and thus keep the animal from being swept under the wheels and possibly derailing the locomotive .The point-rods worked too well. The very first animal *John Bull* hit was, you might have guessed, a large bull, and the heavy creature was speared so securely that it was almost impossible to get him off the prongs. Dripps quickly substituted, first a straight iron bar, and then two bars that came to a not-too-sharp point, resulting in a pilot or cowcatcher much as we know it today.

Locomotive builders began to supply a bell as standard equipment around 1840, sometime before cabs became standard—as a builder's drawing of the Baltimore & Ohio's mile-a-minute *Mercury* shows it with a bell. Bells were necessary for the same reason that cowcatchers were, for safety—the bell to warn people who were crossing the tracks at grade crossings, and trespassers, both animal and human.

Locomotive headlights were another necessary American innovation; even today many European locomotives do not use them—their only lights consisting of marker lights indicating the kind of train the engine is pulling. At first, no American trains were run after dark, since most of the early railroads were so short that all trips could be made during daylight. The first attempt to light up the track ahead of the locomotive was made by Horatio Allen on the 135-mile South Carolina Canal & Railroad Company route about 1833. Because of its length it was difficult to avoid night operation, so Allen placed two small flat cars ahead of a locomotive and built a bonfire of pitch-pine on a pile of sand on the leading car.

The first practical and enclosed headlight maker has been

lost somewhere along history's trail but Baldwin's *Alert*, built in 1837, sported, as we have mentioned, a headlight—but whether the headlight was added later we do not know. In 1840 the new Boston & Worcester Railroad fitted a locomotive with a real headlight, with a reflector behind the animal-oil soaked wick. By the 1850's a box-shaped sheet-iron headlight containing a tin reflector and a whale-oil lamp was in fairly general use. On the more "fancy" locomotives the headlights were mounted on ornate brass-scroll brackets, and were highly decorated with paintings of animals, landscapes and even portraits.

The locomotive whistle, whose rich minor chords have been heard by generations of Americans with emotions that ranged from nostalgia to wonderlust, originated, unlike most of the "extras" adorning American locomotives, in England. In 1833, after a crossing accident on the Leicester & Swannington, a device accurately known as a "steam trumpet" was applied to the locomotive *Comet* to be used as a warning signal. The idea took hold and soon the steam whistles, which had been used on stationary steam engines in England for several years, began showing up on locomotives.

The first American locomotive to be equipped with a whistle was probably the *Sandusky*, built by Rogers, Ketcham and Grosvenor of Paterson, New Jersey. This horizontal-boilered 4-2-0 type engine, built in 1837, was to become the first locomotive in the state of Ohio, but was first tested on the Paterson & Hudson River Railroad .The day the test was made, J. H. James, president of the Mad River & Lake Erie Railroad, to whom the locomotive was consigned, was far more interested in shocking the assembled crowd with the whistle's shriek than in the locomotive itself and tooted the engine's warning signal so long and often that the *Sandusky* ran out of steam several times and so made a poor operational showing.

The whistle fitted to the *Sandusky* was a high pitched single-note instrument that gave out a shrill screech much unlike the mellow chords associated with the locomotives of today. The shrieking single-tone whistle was not, however, replaced by the lower pitched "chime whistle" which combined several notes into a melodious chord until the 1880's.

Whistles soon became much more than a warning to get off the track. Using a standard code of long and short blasts the engineer could signal or acknowledge signals with the train crews, either his own crew or that of another train. Another, and unofficial use of whistles, was to change their tone by carefully controlling the amount of steam released, and many an engineer used his whistle to signal his wife or sweetheart along his route. Some engineers became quite adept at this art, developed an individual style of whistling, and some could even play a tune! More than a few engineers purchased, at their own expense, the melodious steamboat whistles of their day and installed them on the locomotives assigned the engineer, and took the whistle along each time they were transferred to another engine. Some engineers even purchased their own headlights, being dissatisfied with the drab or inadequate lamp that had been installed as standard equipment on their engines.

Another American locomotive "extra" was the cab. Although the first locomotive built with a shelter for the engine crew was probably the *Samuel P. Ingham*, built for the Beaver Meadow Railroad, by Garrett & Eastwick in 1833, cabs were not standard equipment until several years later. It is really hard to tie down the actual date of the first cab because the first railroads built in New England encountered bitter weather and who is to know the first Yankee engineer or fireman with the ingenuity to knock together a cab. It was common for the crew to lose either fingers, toes, and even noses by frost-bite, so engines came to be fitted with wooden box-like structures with, at first, canvas then later glass windows, a box which, although usually open at the rear, gave plenty of protection, being warmed often beyond the point of comfort by the radiated heat from the fire-box and boiler.

1ST MUD DIGGER TYPE

DESIGNED BY ROSS WINANS

Builder . . . Matthias W. Baldwin & Company . . . 1844

2ND MUD DIGGER TYPE

HORIZONTAL-BOILERED FREIGHT AND PUSHER ENGINE

Builder . . . Ross Winans, Baltimore . . . 1844

1ST MOGUL TYPE ENGINE

HEAVY FREIGHT BUILT FOR RUSSIA

Builder . . . Eastwick & Harrison . . . 1842

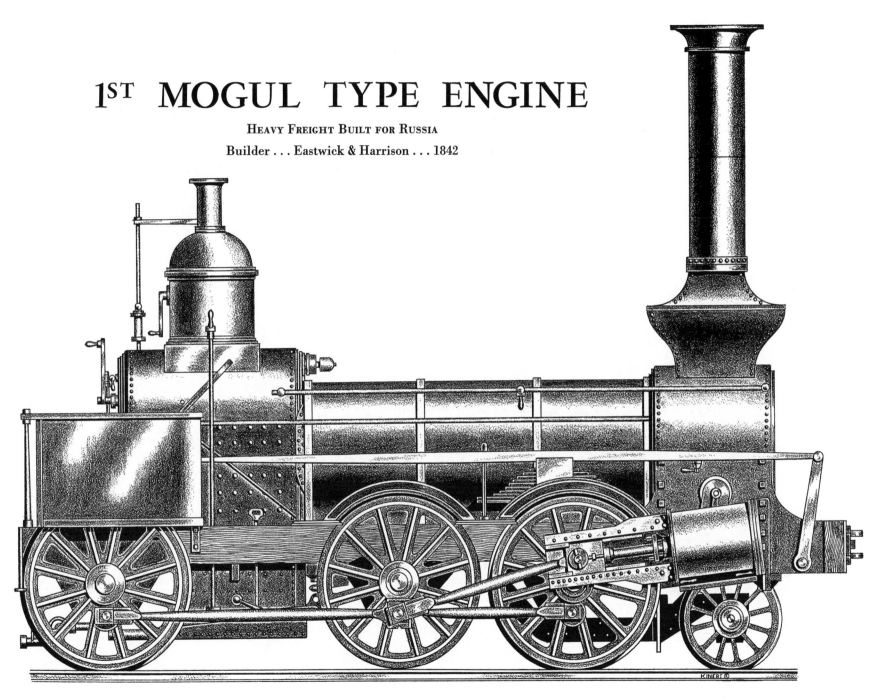

CHESAPEAKE

10-Wheel Freight Engine

Builder . . . Richard Norris & Son . . . 1847

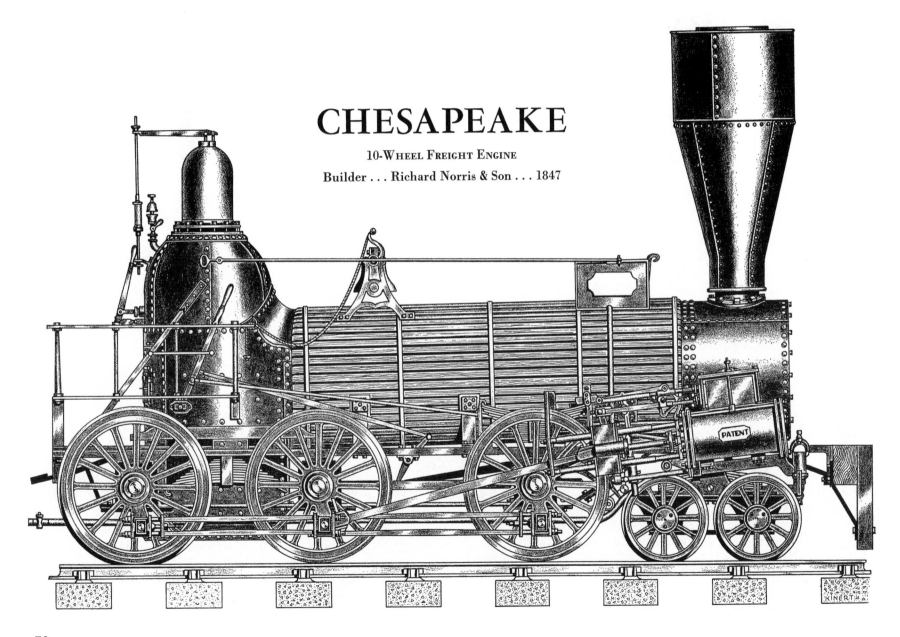

CHAPTER X

NEW RAILS AND LARGER ENGINES

Almost all early railroads used heavy blocks of granite as their roadbed, the blocks set in parallel rows, leaving the center lane uncluttered for horses hoofs. Long wooden stringers, topped by narrow strips of iron were set on the granite blocks and affixed thereto by spikes. The iron strips frequently pulled loose from the wooden stringers because of the weight of the locomotive passing over it, then the torn iron would spear through the following passenger car floors, sometimes impaling the passengers .This kind of track, while not too unsatisfactory for horse-drawn cars, was entirely too poor for steam trains. The granite blocks would shift when it rained and crack in freezing weather, making it difficult to keep the track level and in line. Then, when the track became uneven the unyielding granite blocks caused a pounding effect upon locomotives and cars, causing them to wear out quickly and causing the customers a very hard ride in the unsprung coaches of that day.

As we mentioned earlier, Robert Stevens, Chief Engineer of the Camden & Amboy Railroad, journeyed to England, in 1830, to purchase locomotives for his railroad. While on the sailing ship which took him across the Atlantic he whittled a T-rail design out of a block of wood and, after reaching England, contracted with ironmasters there to roll him some iron T-rails .The rails were made, shipped across the ocean and laid on the Camden & Amboy roadbed in 1831. They proved to be a great improvement on the strap rails then in use and so rails of the same T design soon came into general use, though, because of their high price, a few American railroads continued with strap rails for several years.

An inverted U type iron rail was tried by the B & O and other railroads but it was soon found that the T-rail was superior. This basic shape is, of course, in use to this day.

The early miles of the Camden & Amboy had its rails laid on the typical granite piers of the day. The granite was quarried at Sing Sing, New York, and the quarrymen fell far behind in their deliveries so, as a temporary expedient, Stevens had his men lay a bed of wooden cross-ties. Both to his surprise and that of other pioneer railroad men, the cheaper track foundation gave a smoother ride to the train, permitting much easier replacement and cheaper maintenance and, what is more, it solved the problem of the rails becoming too far apart or too close together as in the use of ever shifting granite blocks. So the modern roadbed was born—and it was only natural that progress demanded faster and more powerful engines to go along with the new rails and roadbed.

The locomotive *Experiment* had been a big step forward by allowing the locomotive to guide itself down the track with its leading four-wheel truck but it seemed the practical size for engines had been reached. Then, in 1836, Henry R. Campbell, chief engineer of the Philadelphia, Germantown & Norristown Railroad, took another big step forward when he had James Brooks of Philadelphia build a locomotive to his own patented design, with a four-wheel leading truck and four drivers, with one pair of driving wheels in front of the firebox and another pair behind, thus creating a new type engine which became known as the American 4-4-0 type. In designing the new engine Campbell desired to produce a locomotive that would be easy on the fragile track of its day, even though more powerful, strap rails still being the rule on most roads.

Campbell knew that his engine was a new type but he could not have possibly foreseen that the type was to almost completely revolutionize railroading in the United States, that the new type would almost dominate the rails of this country for better than forty years and be copied by almost every railroad in the world. This first American type locomotive was not, in itself, an outstanding success for it was very top heavy and far too rigid because of its massive outside frame and rode the rails hard because equalizing driver weights had not been thought of.

Then, just one year later, a Joseph Harrison, Jr., also of Philadelphia, patented the equalizing beam, by which the weight of the locomotive was suspended at three points—the kingpin of the lead truck and the center of two equalizing beams, each of which rested on the drive-wheel axles. Now all four drivers on the newly invented American type engine, able to move up and down independently of the others, could rest firmly on the rails, despite the roughness of the track, eliminating the slipping and spinning of rigidly attached drive-wheels. Recognizing Harrison's talent, Garrett & Eastwick quickly hired him as their foreman.

The first of the true "equalized" American types was the *Hercules*, built by Garrett & Eastwick of Philadelphia for the Beaver Meadow Railroad. The *Hercules* of 1837 was so flexible that it would accomplish more work than other engines in use so more like it were ordered. Like other locomotives of her day, the *Hercules* was without cowcatcher or cab and had no whistle but was one of the first engines to sport a bell and, what is more important, did have the 4-4-0 wheel arrangement.

The Camden & Amboy Railroad exercised a powerful influence on the development of the locomotive engine, due principally to the ability of Robert L. Stevens, president of the line, and Isaac Dripps, a young mechanical genius. The first rails of the railroad were laid at Bordentown December 4, 1830. While abroad purchasing T-rails for his new road Stevens saw Stephenson's *Planet* type engine in its trials so ordered one like it for his road. It was named *John Bull No. 1*, was completed in May 1831, and arrived at Bordentown in August by boat. The first work which Mr. Dripps did for the railroad was to assemble the *John Bull*. He had never seen a locomotive before but studied the mechanism carefully and erected it without a mistake! The *John Bull*, an 0-4-0 inside connected engine with the frame outside its wheels, became, under Dripps expert care, the first successful English built locomotive in America and still exists today.

Although Mr. Stevens ordered several new engines from England he did not propose to be dependent on foreign makes so opened shops in Hoboken for the construction of locomotives, with Mr. Dripps in charge. In the first few engines he built, Dripps imitated the *John Bull* wheel arrangement but designed a two-wheel lead truck to carry some of the weight off the front driving wheels and found that it also helped steer the engine down the track. Also, as we mentioned earlier, the pilot also became the first cow catcher.

The opening of the Camden & Amboy Railroad greatly stimulated traffic between New York and Philadelphia so Stevens and Dripps planned to build engines more powerful than anything previously thought of. They first designed a class of locomotive that became known as the *Monster* and the first engine was assembled at Bordentown in 1836. Built as an anthracite burning freight engine the boiler of the first *Monster* was deficient in steam making and after being in use some time blew off the dome.

The engine part of the *Monster* was very peculiar, being influenced by Stevens and Dripps previous connection with steamboat engines which then included many curiosities. The *Monster* was an eight-wheel connected engine, each set of four wheels being free to move in a plane slightly different from the other set which gave it flexibility in passing curves, an

LIGHTNING

SINGLE-DRIVER FAST PASSENGER ENGINE

Builder . . . Edward S. Norris Works . . . 1849

idea following Allen's double-ended locomotive built for the South Carolina Railroad.

The cylinders were bolted to the boiler sides at a 30 degree angle and the piston worked towards the front. Each set of wheels had connecting rods and there was no frame, all attachments being fastened to the boiler itself! The *Monster* type was not a success but the original engine and three others built at the Trenton Locomotive Works were far ahead of their time in many ways, and were relatively more powerful than any locomotives built up to 1910. Rebuilt as 4-6-0 types they performed yeoman duty for years.

United States designers and builders, with Philadelphia as the hub of activity, were, with true Yankee ingenuity, cribbing each others designs, improving on them and adding new engine types almost weekly as America laid her rails to the north and to the south as well as ever westward towards the boat traffic of the Ohio and Mississippi Rivers.

In the summer of 1839 Eastwick & Harrison, the latter having been made a partner, received an order from the Philadelphia & Reading Railway for a freight engine which was to burn anthracite coal in a horizontal tubular boiler. It was not to weigh more than 11 tons, 9 of which were to rest on the driving wheels. Named the *Gowan & Marx*, the engine was designed on the plan of the *Hercules* and, to distribute its weight, the rear axle was placed under the firebox. A blower of the steam-jet type was used for the first time, used as an aid for more draft for the furnace.

When put in service, the *Gowan & Marx* proved to be extraordinarily powerful. On February 20, 1840, it hauled a train of 101 loaded four-wheeled cars from Reading to Phila-delphia and averaged 9.82 miles an hour. The train's gross load was 423 tons and, if the weight of the engine and tender is included, the total was 40 times the engine's weight. This was believed to excel by far any performance on record for an 11-ton locomotive. Its great success led the Philadelphia & Reading to order ten more engines of its type.

William Norris, in 1838, had come out with a cabless 4-6-0 freight engine that quickly won success by pulling heavy freight for the City Point Railroad through mountainous Virginia. And in the same year Norris built a 4-4-0 American type engine which sported a headlight and cab. Then, in 1842, both Norris and Baldwin built 6-coupled engines, 0-6-0 wheel arrangements with all six wheels connected by rods to deliver drive power.

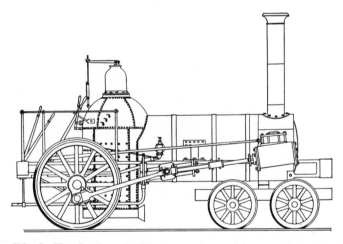

The *Black Hawk*, first Baldwin engine with outside cylinders. Built for the Philadelphia & Trenton Railroad by Baldwin in 1835.

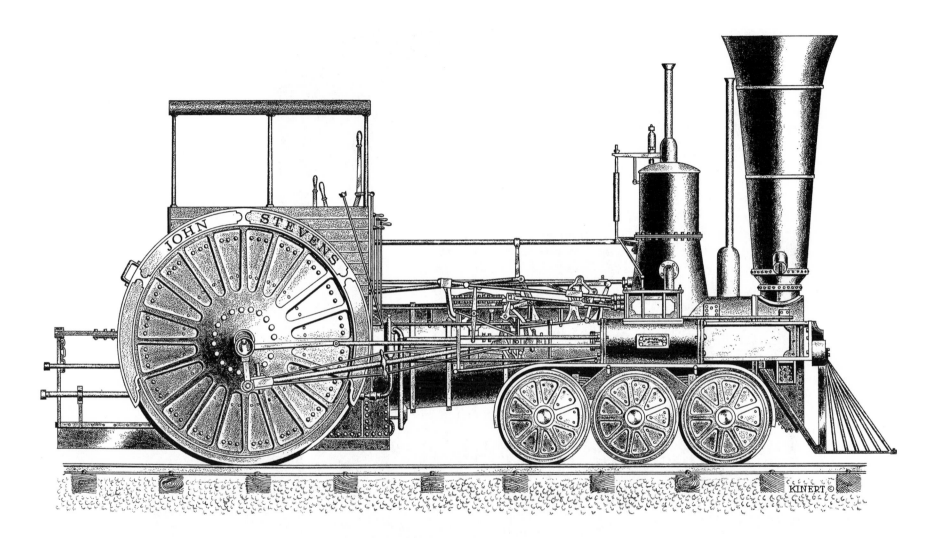

JOHN STEVENS

Fast Passenger Engine — Designed by Isaac Dripps

Builder . . . Richard Norris & Son, Philadelphia . . . 1849

An excellent early steel engraving depicting a varient of the STEVENS type locomotives built for the CAMDEN & AMBOY RAILROAD. Note the "winterized" cab. The odd tender, typical of those used by the C. & A. R.R. at this period, served as a combination fuel carrier and caboose for train crew members.

CHAPTER XI

MUD DIGGERS, CAMELS AND HIGH DRIVERS

Down in Baltimore town Ross Winans was also busy with new ideas and, after building his "Crab" engines for the B & O, designed an eight-wheel coupled locomotive, three of which were built by Matthias Baldwin in 1884 for the Western Railroad of Massachusetts. The trainmen promptly nicknamed this engine type "Mud Digger" because of the odd appearing linkage movements when the engine was working. The engines had horizontal cylinders like the "Crabs" and the upright boiler was still retained. This eight-wheeler was practically a double truck locomotive. It ran steadier than the four-wheelers and made good power for hauling heavy trains, but no more of this type was built for horizontal-boilered locomotives were in demand.

Winans led the world in advocating powerful locomotives and had a clear conception of the economy that would result using engines as large as the track would carry. The light track on which his engines had to run kept down the weight some but he built engines that compared fairly with engines of 1905.

In late 1844 Winans designed and built another and more advanced "Mud Digger" type, the first 0-8-0 engines, those having eight-coupled wheels, to be set on B & O rails. They were somewhat similar in cylinder and drive arrangement to his earlier "Crab" and first "Mud Digger" type but this new engine was much larger and was the first Winans engine to have a horizontal boiler.

The main connecting rods on this second "Mud Digger" type, were coupled to cranks on a shaft extending across the frames in the rear of the fire box and geared by spur wheels to the back driving axle. The driving axles carried end cranks which were coupled by side rods. As the main and side rods moved in opposite directions, by reason of the interposed gearing, the engine presented an odd appearance when running. This odd machinery, both coming and going, also churned up dust and loose dirt from the track so deserved the nickname "Mud Digger" even more so than the earlier type, but each of the first 12 engines built had its own name. That illustrated, No. 37, is the *Cumberland* as it appeared in 1863, complete to headlight and cab.

The horizontal-boilered "Mud Diggers" were built at different dates, from October 1844 to December 1846. Their performance in service was exceptional and some were in use as late as 1865 or after.

The class of engine which followed the "Mud Diggers" in the B & O's constant search for more power was built by Matthias Baldwin. It had inclined cylinders and what came to be known as the Baldwin "flexible beam truck," which carried the bearings of the front and second driving axles. An eight-wheel connected type, the first of these, the *Dragon*, No. 51, was placed on the road in January 1848 and four more engines of this type were built by Baldwin. The *Memnon*, No. 57, shown in the photo section, and at least one other of this class, was built by the Newcastle Manufacturing Company and the B & O shops also built three of the eight-wheelers, then no more of this class were built for B & O rails.

Meanwhile, Ross Winans had not been idle down at his Baltimore shops. Quick to realize the inherent good features of the American 4-4-0 locomotives, Ross built three of the type, but not one to follow in the footsteps of others he startled the designers of 1848 by designing and building a radical and distinctive class engine of 0-6-0 wheel arrangement, the now famous "Camel" type. It was the last great invention from his fruitful brain.

Designed to pull the ever increasing loads over the Alleghenies, these freight engines were perhaps the most curious large engines ever run on an American railroad. Winans placed the cab atop the boiler's "hump" for better viewing of the mountain curves ahead and at the same time gave the engineer and fireman fresh air to breathe in the numerous tunnels along the line, for the smoke trailed harmlessly over the cab's roof. It was this placing of the cab that earned for the first of these engines the name *Camel*, engine No. 55. Because the fireman fueled the furnace at the rear of what looked like a huge cannon the engines were also called "muzzle loaders."

Built in Winans' shops for the Baltimore & Ohio Railroad, the camels were made in three classes, the "short," "medium," and "long" furnace models. All had cylinders 19"x22" and 43" driving wheels except the first camel, No. 55, which was placed on the road in June 1848. Its cylinders were 17"x22" and was probably the only camel of the "short" furnace type and the only one with the smaller cylinders.

The camels marked a great advance in freight motive power. More than 200 were turned out of the Winans' shops by 1863, 120 of these being for the B & O alone. They were also used on the Pennsylvania, Philadelphia & Reading, New York & Erie, and other roads.

Soon after delivering No. 219, built in February 1857, to the B & O, a controversy arose between Ross Winans and Henry Tyson, then master of machinery for the road, as to the relative merits of the camel and a newer concept locomotive, the six-coupled ten-wheel engine, the result being a cessation of orders by the B & O.

Winans soon quit building locomotives in disgust. He sided, though passively, with the South in the Civil War and this too may have influenced him.

Samuel J. Hayes, who was master of machinery for the B & O Railroad from late 1851 to the spring of 1856, designed and built in 1854, in the company shops, the first of several engines that followed generally the lines of Winans' camels, but differed principally in having a four-wheel leading truck, larger driving wheels and several other improvements. They were known as the "Hayes ten-wheelers" and were efficient and satisfactory both in passenger service, for which they were originally intended, and for moving freight. A number were built in the company's shops, 17 being constructed up to 1860 alone. Others were built by various firms for the B & O. In 1901 the last ten-wheel camel then in service, No. 173, traveled under its own power to Purdue University, where it is now preserved with other old locomotives.

The camel concept locomotive was built in various forms by several firms and railroads as 0-6-0 and 4-6-0 freight types and 4-4-2 passenger engines, many being built long after the turn of the century.

In 1845, about the time Ross Winans was concerned with his second "Mud Digger" type freight engine, down at Baltimore Isaac Dripps had set about designing another locomotive for the Camden & Amboy Railroad, this one for passenger service. The plans were approved in 1847 and the order placed with Richard Norris & Son of Philadelphia. The first engine, named the *John Stevens*, No. 28, was first tried on April 17, 1849. The engine had six small lead wheels and only two drivewheels which were 8 feet in diameter! After the engine had been in service awhile Dripps realized that the over-sized single pair of drivers did not have sufficient adhesion to haul the passenger trains then operated. The six-wheel truck was pivoted so far back on the boiler that it carried most of the engine weight, making the drivers very deficient in adhesion and prone

LIGHT PASSENGER ENGINE

Builder . . . Niles Locomotive Works, Cincinnati . . . 1852

to get off the track. Dripps was over-ruled, however, and several more were built, some with smaller driving wheels.

The driving and truck wheels were of wrought iron and the spaces between the spokes were filled with wood, either an early attempt at streamlining or simply to dress up the looks of the engines. The fire door was below and behind the axle and the fireman stood in a pit, the bottom of which was on a level with the bottom of the ash pit. On top of the dome was a safety valve whose lever extended to a spring scale to show the pressure carried, steam gauges not then in use. The safety valve was encased to prevent tampering with it!

No sandboxes were provided and, with little weight on the drivers the engines were very slippery and slow to get a train under way. And the boiler was not large enough to furnish ample steam for the large cylinders. These engines ordinarily hauled a train of six four-wheeled passenger cars but this was more than they could handle well. Once in motion, with a light train, they could run as fast as anyone would care to travel! These engines ran from 1849 to 1862, one being in service as late as 1865, but most of them had been rebuilt as 4-4-0 types by then.. .

Early locomotive designers the world over made the common mistake of believing that the size of driving wheels, instead of the boiler, controlled the speed capacity of a locomotive, and American designers went along with this fallacy. The *Stevens* was the first U.S. engine conspicuous for its great size driving wheels. In 1849, the year the *Stevens* was put to work, Edward S. Norris of the Schenectady Locomotive Works, built the locomotive *Lightning* on order for the Utica & Schenectady Railroad, with a single pair of driving wheels 7 feet in diameter. The *Lightning*, also a 6-2-0 type, ran only about a year as it proved to be a failure.

Several early railroads had locomotives built with a single pair of 7-foot drivers, which seemed the standard size in those days. The *Stevens, Lightning,* and the *Indiana,* the latter being one of three of this type built for the Pennsylvania, are the best known examples. All three engines are illustrated in these pages. Passenger locomotives with 7-foot driving wheels later became quite common but there were more than two drivers and they had much more weight on the drivers. They also had ample power to drive them, with sufficient boiler capacity to provide all the steam required.

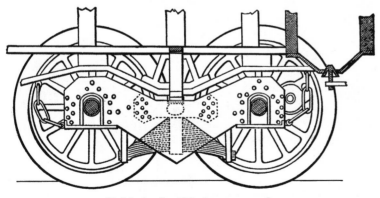

Baldwin flexible bearer truck

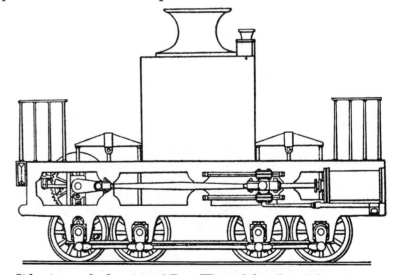

Side view scale drawing of Ross Winans' first "mud digger" type

An exquisitely detailed and operative replica of the *Best Friend of Charleston*, built from working drawings by the Southern Railway.

Credit: *Southern Railway*

Beautiful and exact replica of the *DeWitt Clinton,* first steam-powered passenger engine in New York State. Built in 1920, the *DeWitt* is housed in the Henry Ford Museum at Dearborn, Michigan.

Credit: *NYC*

82

The locomotive *Mississippi* is the South's oldest existing railway engine. Built in 1834 it apparently first saw service on the Mississippi Railroad, which ran northeasterly from Natchez to Hamburg, Mississippi, a distance of about 19 miles. This diminutive little engine worked for several railroads until about 1890, was reconditioned and ran under her own power as late as 1925. The tender shown is not the original. On display at the Museum of Science and Industry, Chicago, Illinois, the *Mississippi*, a wood burner, weighs 7 tons without tender. Diameter of drivers, 43 inches, tractive power, 4,821 pounds.

Credit: *ICR*

Davis and Gartner, who built the *Atlantic* in 1832, built as their next two "grasshoppers" the *Traveller* and the *Arabian*. Neither of these locomotives are extant but, thanks to photography which came into commercial use around 1850, this fine shot of the *Arabian* was taken at the Mt. Clare roundhouse. Engineer William Galloway poses in the cab, bedecked in his finest. Water tank and bell are later additions.

Credit: *B&O*

The *Andrew Jackson*, built in 1836 by George Gillingham & Ross Winans at Baltimore, is shown here with the side rod removed that connected the two axles, thus altered to resemble the first "grasshopper" engine built, Davis' *Atlantic* of 1832. The *Andrew Jackson* was B & O No. 7 for many years, later became No. 2. Metal water tank shown replaced original wooden barrels.

Credit: *B&O*

The *John Hancock*, built in 1836 by Gillingham & Winans at Baltimore. Of the many "grasshopper" type engines constructed, three have survived. The *John Quincy Adams*, built in July 1835, is on exhibit in Carillon Park, Dayton, Ohio, while the *Andrew Jackson* and the *John Hancock* (above), both built in 1836, are housed in the B & O Museum at Baltimore. All three engines were in use as switchers at Baltimore until 1892! Metal water tank and bell are not original, two barrels originally supplying water to the boiler.

Credit: *B&O*

Original Baltimore & Ohio *Memnon* 0-8-0 locomotive built by the New Castle Manufacturing Company on sub-contract from the Baldwin Locomotive Works in 1848. Because of the valuable service during the Civil War in hauling supplies and soldiers, it was nicknamed "Old War Horse." The cow-catcher was added in later years.

Credit: *B&O*

The *Pioneer*, one of America's oldest locomotives still to exist, was built in 1851 by Seth Wilmarth in Boston for use as a light passenger engine by the Cumberland Valley Railroad, now part of the Pennsylvania Railroad. Photo taken at the Chicago Railroad Fair of 1949.

Credit: *Gerald M. Best*

The *Daniel Nason,* an inside connected 4-4-0 American type engine built at the Boston & Providence Railroad Shops in 1854. These engines earned the nickname "Dutch Wagons" because of their absence of visible running gear, their more than ample running-board area and the foreign or odd appearance in general. Note the six-wheel arrangement of the tender. Photo taken at the New York Fair Railroad Exhibit in 1939.

Credit: *Gerald M. Best*

The famous locomotive *General*, built by the Rogers Locomotive Works in 1855 for the Western & Atlantic Railroad, as she looked at the Chicago Railroad Fair in 1949. The *General* has resided on display at the Union Station in Chattanooga, Tennessee, for many years, is now being reworked to once again steam on her own.

Credit: *Gerald M. Best*

The gracefully proportioned *Wm. Mason,* a 4-4-0 type built by William Mason of Taunton, Mass., for the B & O Railroad in 1856. Still able to steam under her own power, the *Mason* resides in the B & O Museum at Baltimore. Mason has often been named "Father of the American type engine" because he introduced the revolutionary idea of turning out products of beauty as well as of utility.

Credit: *Gerald M. Best*

The beautifully proportioned and ornate *William Crooks*, built by Smith & Jackson of Paterson, New Jersey, in 1861, was the first locomotive in Minnesota. Brought to St. Paul on a Mississippi River steamboat, the colorful "old No. 1" was set on rails of the St. Paul and Pacific Railroad (now Great Northern Railway) and made its maiden run on June 28, 1862, between St. Paul and St. Anthony (now Minneapolis). The original engine is now on proud display in the St. Paul Union Depot.

Credit: *Gerald M. Best*

This exceptionally fine locomotive honors its builder Thatcher Perkins, who designed and supervised construction in 1863 at the B & O's Mount Clare Shops in Baltimore. The first 10-wheel (4-6-0) type passenger engine built, she was designed for moving first-class passenger trains over the steep and winding grades of the Allegheny Mountains between Cumberland, Maryland, and Grafton, West Virginia. In addition to its sturdy construction and powerful lines, the engine has symmetry and grace and with its decorative colors presents an unusually attractive appearance. Shown here at the Rail Fair in 1949 she resides today in the B & O Museum in Baltimore.

Credit: *Gerald M. Best*

The *Reuben Wells* locomotive, which was designed by Reuben Wells, master mechanic of the Jeffersonville, Madison & Indianapolis Railroad, now part of the Pennsylvania Railroad System. The *Reuben Wells* was placed in service August, 1868, on the steepest inclined railroad track in the world, extending from Madison, Indiana, to North Madison. Shown here at the Chicago Railroad Fair in 1949 the *Reuben Wells* is still in excellent mechanical condition, can run under her own power. The caboose shown is of much later vintage.

Credit: *Gerald M. Best*

An unusual 4-4-2T type built by R. Norris & Son in 1868 as Central Pacific Railroad No. 1. When the Southern Pacific Railroad acquired Central Pacific the engine became S.P. No. 1003, later worked as No. 1 for the Sacramento Shops as a switcher. Note the link and pin coupler on the pilot.

Credit: *SP*

The *Minnetonka* 0-4-0T (tank over boiler), built by Smith & Porter, Pittsburgh, Pennsylvania, was purchased by the Northern Pacific on July 18, 1870, and was used on early railroad construction work in Minnesota. Still able to steam on her own, the *Minnetonka* is pictured at the Chicago Railroad Fair in 1949.

Credit: *Gerald M. Best*

An 0-4-2T built in 1871 as No. 1 engine for the Tuskeegee Railroad by the Danforth Locomotive & Machine Company of Paterson, New Jersey, their No. 740. Photo taken at Atlanta, Georgia, in July, 1906, after the Southern Iron & Equipment Company had reworked the engine for sale to the Lane Brothers Company of Esmont, Virginia, their engine No. 7.

Credit: *Gerald M. Best*

A nicely proportioned American type (4-4-0) locomotive built by the Central Pacific in their shops at Sacramento in 1872. Although most western railroads continued through the years to purchase their locomotives from the East the C.P.R.R. built several of their own in the 1870's and 80's.

Credit: *SP*

The *J. W. Bowker*, a 2-4-0 type locomotive, was built in 1875 by the Baldwin Locomotive Works, Philadelphia, for the Virginia & Truckee Railroad. Retired, it has been acquired by the Railway & Locomotive Historical Society (Pacific Coast Chapter).

Credit: *Gerald M. Best*

The *J. C. Davis*, built in the B & O Mt. Clare shops in 1875, was the first passenger Mogul (2-6-0) locomotive in the world. It won first honors at the Centennial Exposition in Philadelphia in 1876 as the finest and largest locomotive built up to that time. The *Davis*, designed for heavy passenger service over the Allegheny Mountains, reposes in the B & O Museum at Baltimore. Length of engine, 40′ 2½″; height, 14′ 8½″; total length, with tender, 58′ 2¼″. Drivers, 56″ diameter; total weight on drivers, 76,550 lbs. Engine weight, 90,400 lbs. Tractive power, 8,580 lbs.

Credit: *B&O*

Fulton County Railway No. 1, a 4-4-0 American type engine built by Baldwin in 1879. The Railway was 3 ft. narrow gauge. Photo taken at Lewistown, Illinois, in 1904 when the little engine was still hard at work.

Credit: *Gerald M. Best*

Illinois Central suburban locomotive No. 1402, a 2-4-4TB type, the TB designating tender back, that the tender was built on the same frame as the engine itself. This particular type, built by the Rogers Locomotive Works in 1880, was nicknamed "teakettle" by Central employees. Photo taken at Chicago in 1926.

Credit: *Gerald M. Best*

An interesting 0-4-0T switcher built by the Hinkley Locomotive Company of Boston, Massachusetts, in 1880 for the Galveston, Harrisburg & San Antonio Railroad as their No. 42 engine. It later worked as No. 7 for the Texas and New Orleans Railroad, a subsidiary of the Southern Pacific. Photo taken at Houston, Texas, in 1924 where the engine worked as a shop switcher.

Credit: *Gerald M. Best*

This 0-4-0T, built in 1881, was the 457th engine turned out by the Pittsburgh Locomotive Works and was sold to the Stacy Mining Company as their No. 1 engine. Photo was taken September 26, 1907, after the engine was reconditioned by the Southern Iron & Equipment Company (their No. 569) of Atlanta, Georgia, a firm which purchased used locomotives and rolling stock by the hundreds and reworked them for resale.

Credit: *Gerald M. Best*

This well-proportioned 3 ft. narrow gauge 2-6-0 Mogul type engine was built by Baldwin in 1881 for the Nevada Central Railroad as their No. 2. It worked until 1938 and helped the salvage crew tear up its own track. Purchased and reworked by Ward Kimball of Arcadia, California, it now resides in Ward's backyard roundhouse and steams over his privately owned Grizzly Flats Railroad as desired. Photo taken in 1956.

Credit: *Gerald M. Best*

The Virginia & Truckee American type (4-4-0) *Genoa,* built in early 1880's repainted to play the part of *Jupiter* at the Chicago Railroad Fair in 1949. *Jupiter* was the Central Pacific engine that met Union Pacific No. 119 at Promontory, Utah, on May 10, 1869, to join the continent by rails. The *Genoa* is still a wood burner and can run under her own power.

Credit: *Gerald M. Best*

The *Olomana,* a 3 ft. narrow gauge 0-4-2T (water tank over boiler) type engine, was built by Baldwin in 1883 for the Waimanalo Sugar Company of Oahu, Hawaii, and hauled sugar cane into the mills until 1947, when the plantation was sold for a housing development. Purchased in 1948 and reworked into like new condition by Gerald M. Best, the engine resides today in Arcadia, California. Photo taken in 1958.

Credit: *Gerald M. Best*

This superb little 3 ft. narrow gauge 0-4-2TB (tank back of engine on same frame) was built by Baldwin in 1883 for the Waimanalo Sugar Company of Hawaii, worked until 1947. Purchased and restored in 1948 by Ward Kimball (in cab) of Arcadia, California, *Chloe* resides in Ward's privately owned roundhouse and runs on his shortline whenever called upon. Photo taken in 1956.

Credit: *Gerald M. Best*

This 2-8-0 freight engine was built by Baldwin, their No. 6713, in 1883 for the Western New York and Pennsylvania Railroad, their No. 42, and later worked for the Pennsylvania Railroad as their engine No. 6282. Photo was taken at Atlanta, Georgia, in June, 1910, after the engine had been reconditioned for resale by the Southern Iron & Equipment Company, their No. 667.

Credit: *Gerald M. Best*

Willie, an 0-6-2T switch engine built by the Dickson Locomotive Works, Scranton, Pennsylvania (their engine No. 449), in 1883 for the Chicago, Fairchild and Eau Claire River Railroad. This is a fine example of ornate and precision workmanship combined with utility. *Willie* was worked over 3 ft. narrow gauge track.

Credit: *Gerald M. Best*

The beautifully ornate Lehigh Valley Inspection engine *Dorothy,* built at the Wilkes-Barre Shops in 1884, was used to haul the super-intendent and other officials over the road. Photo was taken at Easton, Pennsylvania, in 1895. It is said that this engine, after its retire-ment, was purchased by a Lehigh Valley Railroad employee, in whose backyard it rested for many years. The employee has since died but the locomotive may still exist.

Credit: *Gerald M. Best*

The *Asa H. Curtis,* a 3 ft. narrow gauge 2-6-0 locomotive built in 1885 by the Brooks Locomotive Works of Dunkirk, N. Y., for Henry Stevens & Company, a lumbering concern in Michigan. Photo is a builders print, taken at the Brooks Works.

Credit: *Gerald M. Best*

Atlantic Coast Line locomotive No. 3, an American type built by Rogers, their No. 3572, in September, 1885. Sparks can plainly be seen emitting from the stack of this old wood burner. Note the link and pin coupler on the pilot, a device still used at that late date. Drive wheel diameter 62″, cylinders 18″x24″. Total weight of engine 90,000 lbs., tender weight 62,000 lbs.

Credit: *ACL*

This nicely proportioned 4-4-0 American type engine No. 374 was built by the Canadian Pacific Railway in their shops, their No. 1038, in 1886 and was presented to the City of Vancouver, B.C., in 1945 for display, at which time this photo was taken.

Credit: *CPR*

An American type 4-4-0 locomotive built by the Brooks Locomotive Works, Dunkirk, New York, in 1887 for the Ulster & Delaware Railroad, their No. 3 engine. Photo taken by the builder at the factory.

Credit: *Gerald M. Best*

This coal-burning locomotive was designed and built by A. J. Cromwell, master of machinery at Mt. Clare, Baltimore, in 1888 and was named after him. On display at the B & O Museum, this engine was the first B & O Consolidation type (2-8-0) to use double brick arches in the firebox for better steam generation.

Credit: *B&O*

This 4-6-0 type freight engine was built in 1888 by the Schenectady Locomotive Works for the Northern Railway of California as their No. 1010. When the Southern Pacific absorbed the Northern Railway the engine was renumbered 2174 and worked until 1948 when she was scrapped.

Credit: *Gerald M. Best*

An 0-4-0T built in October 1889 by the H. K. Porter Company, their No. 1080, for the Bailey Libby Company of Charleston, South Carolina, who operated a 3 ft. narrow gauge road. Photo was taken at Atlanta, Georgia, in 1909 after the engine had been overhauled for resale by the Southern Iron & Equipment Company. The H. K. Porter Company of Pittsburgh built many steam locomotives, specialized in small tank over-boiler types of both narrow and standard gauge for use as industrial switchers and on mining roads.

Credit: *Gerald M. Best*

Illinois Central suburban Chicago engine No. 1447, a 2-6-4TB type, built by the Schenectady Locomotive Works in 1889. This engine and other steam types were used in suburban commuter service for many years until being replaced by electric interurbans. Photo taken at Chicago, 1926.

Credit: *Gerald M. Best*

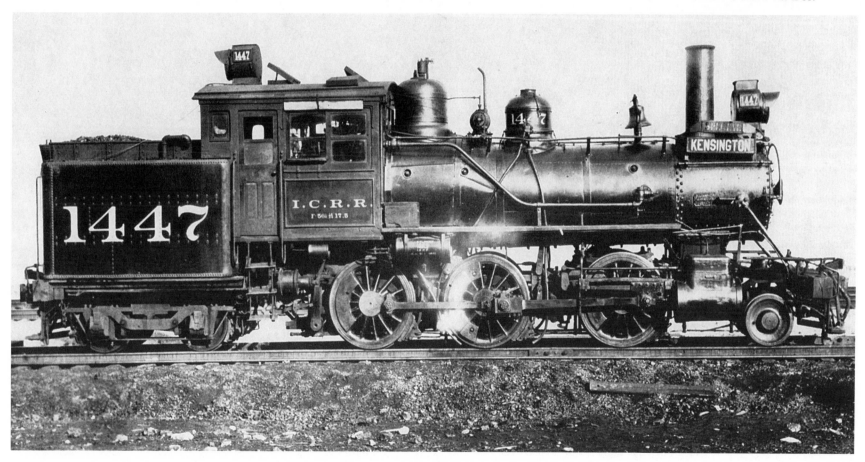

Baldwin-built locomotive No. 10798, a 2-6-0 type built in 1890 as No. 1 engine for the Richmond, Fredricksburg & Potomac Railroad who later sold the engine to the Georgia & Florida Railroad who renumbered it No. 42. It was later purchased by the Southern Iron & Equipment Company, reconditioned by them and sold to the Central Limones Sugar Company of Cuba (their engine No. 3), December 31, 1917. Photo taken at Atlanta, Georgia, in 1917.

Credit: *Gerald M. Best*

A 2-6-0 type built by the Brooks Locomotive Works, their No. 1982, in 1891 as No. 2 engine of the Smithsonia & Dunlap Railroad. It was purchased by the Southern Iron & Equipment Company who reworked it for sale to the Deal Lumber Company, their No. 4 engine. Engine was photographed at Atlanta in January, 1918.

Credit: *Gerald M. Best*

An early Climax geared engine, their No. 475, built in 1892. Picture taken at Atlanta, Georgia, in March, 1911, when engine had been refurbished for the Scott-Lambert Lumber Company as their No. 770. Each wheel on this type locomotive was geared from the drive shaft and was thus exceedingly powerful. The engines could be used on very steep grades but top speed rarely exceeded 18 m.p.h.

Credit: *Gerald M. Best*

This fine example of an American 4-4-0 type was built by the Hannibal & St. Joseph Railroad at their Aurora Shops in 1892 as engine No. 66. It became H & St. J. No. 666, then Chicago, Burlington & Quincy R.R. No. 359 in 1904. It was rebuilt at Denver in June, 1932, for exhibition at the Chicago Century of Progress and was lettered Burlington & Missouri River R.R. No. 35. Later it was revamped to its present condition to simulate the previously scrapped Union Pacific No. 119, the engine used at the Golden Spike ceremony when the East and West railroads met to weld the East and West together with bands of steel track.

Credit: *R. P. Morris*

Built as a heavy freight engine by the Schenectady Locomotive Works in 1892 for the Central Pacific Railroad of California this 4-8-0 type was unique in that it was a cross compound engine, with one high and one low pressure cylinder. The left cylinder was smaller but built heavier to take high pressure steam directly from the boiler and that cylinder's exhaust was piped to the low pressure right cylinder. The Southern Pacific Railroad assimulated Central Pacific in 1891, fitted No. 2933 with a newer type boiler in 1895 and reworked her into a simple compound type in 1910, as shown above.

Credit: *Gerald M. Best*

A 4-6-0 diamond stacked "ten-wheeler" built by the Pittsburgh Locomotive Works in 1893 as engine No. 70 for the Wheeling & Lake Erie Railroad, now part of the New York, Chicago & St. Louis Railroad. The engine was renumbered No. 85 in 1897 and was sold by the W&LE in 1917. Photo taken at Huron, Ohio before 1896.

Credit: *NYC & St.L*

This 0-4-4 Forney type engine was built in 1893 by the Rhode Island Locomotive Works of Providence, R. I., for the Lake Street Elevated Railroad, Chicago. Photo was taken at the Southern Iron & Equipment Company, Atlanta, Georgia, in November 1909 after the engine had been reconditioned and sold to the W. D. Kimball Lumber Company, their No. 2 engine.

Credit: *Gerald M. Best*

126

This American 4-4-0 type locomotive was built by the Schenectady Locomotive Works in 1894, worked for the Kansas City Southern Railroad for many years before being dismantled. Cylinders 18x24 inch, drivers 63 inch, boiler pressure 160 lbs.

Credit: *KCS*

The *Claus Spreckels*, a 4-6-0 type built by Baldwin in 1896 as engine No. 1 for the San Francisco & San Joaquin Valley Railroad, now part of the Santa Fe Railroad. It worked for the Santa Fe as No. 309 and was later sold to the Modesto & Empire Traction Company.

Credit: *Santa Fe*

CHAPTER XII

THE WILD IRON HORSE

The T-rails that locomotives of the 1840's and 50's ran on were of rolled iron and, while they were a big improvement over strap rails, and the quality of iron used in the rails was continually improved, they had a bad habit of snapping in cold weather, especially when a fast passenger engine or the heavier freight types imposed too much strain on the brittle metal. Processing of steel for rails was not perfected until 1863 when the Pennsylvania railroad quickly adopted them, becoming the first road to do so.

In addition to troubles with the rails another big drawback to progress on the rails was that cars were still coupled and braked by hand. The first couplings had merely been an iron bar with a hole in each end and a pin inserted in the holes. As cars became heavier after Ross Winans' double-trucked cars came into general use, the link-and-pin coupler, known to railroaders as the "Lincoln" pin, came into general use. This consisted of a slotted wrought-iron drawbar fastened to the end of the cars into which a long oval link was inserted and held in place by an iron pin which was dropped into a hole in the drawbar after the two cars were just exactly in a position for coupling. This link and pin did hold the train together most of the time, but also accounted for the majority of maimed or dead trainmen. This was so because, with link and pin the crewman had to stand between the two cars being coupled in order to steer the link into the socket, then drop the pin that held them together.

The link and pin coupling was the dread of all men who ever had to couple cars. Amputated fingers or hand told all railroad men that this or that man, no matter his present job,

had once been a brakeman on a link and pin road. Most of the locomotives depicted in this book used, on the earlier types, the iron bar and pin type coupler and the later ones the link and pin coupler as the automatic coupler invented by Eli Hamilton Janney was not perfected and patented until the year 1873. The railroads were slow to adopt the more expensive coupler and until about 1890 there were far more link and pins than automatics on American railroads. The maiming and slaughter of trainmen continued.

Another detriment to progress on the rails was the inability of trainmen to slow or stop their thundering charges once they had them rolling. Passenger locomotives were capable of sustained 60 mph speeds or better over relatively straight and level track with a train of four or five coaches and freight locomotives could pull twenty-five or thirty cars with ease by the late 1840's. Railroad engineers resembled Ulysses when he released the winds and raised a tempest that he could not subdue. The means were there to originate the speed, but when a sudden demand arose to stop, the only retarding forces were the natural resistances to the moving mass. An obstruction on the track or a break in the rail ahead was an automatic signal for the engineer and fireman of a heavy or fast moving train to "hit the dirt" and jump to relative safety from the impending wreck—helpless to stop their wood and iron thunderbolts.

From the day it first appeared, the steam locomotive was accorded zealous attention, select talent being devoted to its improvement and careful maintenance. With the train brake it was different for it was not a revenue earner, but a source of expense without profit. It took many years, darkened by disaster, to change that sentiment.

The first locomotives and trains had no brakes whatsoever and it was a common sight to see a train creeping to a scheduled stop, being brought to a halt by everyone on the station platform grabbing on to the train and actually dragging it to rest, while experienced station men shoved wooden rails between the wheel spokes to lock the wheels!

When reversing levers were invented and installed on locomotives it was quickly learned that an engine and its train could be slowed or even stopped by throwing the reverse lever while the locomotive was in motion, providing there was enough clear rail to do so. This method of stopping was satisfactory until trains became heavier and longer.

The next step was to put hand brakes on the tenders. A horizontal level with a bell crank at the end to which was connected a wooden block at the wheel sides was the method used. This type brake can be seen on the *Alert* tender, illustrated. When a call was made for brakes, the fireman dropped the lever out of its notch and either pushed or stood up on it to increase the pressure upon the wheels, while the engineer was busily using the reverse lever.

Trains became heavier and schedules faster so the next step was to install the above braking method on to the platforms of passenger cars. When a call was made for brakes by a whistle signal from the engineer, a brakeman hired for the duty would go about his work just as the fireman did.

Brake shoes were installed and tried on some locomotives themselves but this did not work out as the weight of the whole train and tender was thus imposed on the locomotive, making it prone to be unmanageable and it often led to derailing the engine.

On most railroads the hand lever gave way to a hand wheel and chain. This arrangement can be seen illustrated on many of the locomotive tenders in these pages. When the four-wheeled truck first came into use on railroad cars, the brake attachment consisted of four vertical wooden blocks, two of which were hung between each pair of wheels and were applied by a canting bar fastened between the blocks and operated by a lever connected to a hand wheel by chain. The wooden blocks were lined with leather, which endured wear better than wood and was easily replaced when worn. Later the wooden blocks were replaced by iron shoes.

The first real improvement on the hand brake came when the brakes of both front and back car trucks were connected, so that the hand wheel at one end of the car applied the brakes of both trucks, all eight wheels of the car. Thus a brakeman could apply the brakes to two cars by stepping over the platform between the cars, thus doubling the effectiveness.

This added another hazard to the link and pin operators woes. Many more men lost not only their fingers or hands but their very lives in climbing to the top of pitching, rolling box cars and walking from car to car on wet or icy "catwalks" to twist at the old hand wheel brakes. Passenger coaches enjoyed the luxury of having their hand brake wheels affixed to the platform at each end of the coaches. The now universally used air brakes, invented by George Westinghouse in 1869 did not come into common use until the 1890's, most railroads contending that the cost of changing to air brakes and keeping them in repair, was too great, the same argument they had used against automatic couplers. Once the railroads discovered the happy effects of safety on their business they went wholeheartedly for anything new and better—and the safety of railroads has become almost legendary.

GOV. WILLIAMSON

Cathedral-Windowed Express Engine

Builder . . . Danforth, Cooke & Co. . . . 1853

DUTCH WAGON TYPE

FAST PASSENGER ENGINE

Builder . . . Murry & Hazlehurst Co. . . . 1854

CHAPTER XIII

THE AMERICAN TYPE LOCOMOTIVE

The first really practical steam locomotive to be widely used in the United States came to be called an American or American Standard, a 4-4-0 (four leading wheels, four drivers and no trailing wheels) and is considered the classic locomotive by many students on the subject—and with good reason.

Light and flexible, the American boldly displays the subsidiary parts that contribute to its operation, attached in any convenient place to the exterior of the great boiler that dominates the machine. The cylinders and valve motion that are the heart of the engine's mechanical function are in full view, being somewhat rakishly perched on their own four wheels tucked under the front end of the boiler. Each of the accessories has an outlandish form suggestive of the bravado of a pioneer country. The huge pilot or "cowcatcher" which had to do the job its name suggests, the monstrous stack intended to catch dangerous wood sparks, the steam dome and sand dome or "box," the large headlight and the gaily painted engineer's cab were all given whatever form their functions required, then lightheartedly bolted onto the machine with no apparent concern about the tasteful result of the whole.

The British locomotive of pre-1900 vintage is a more organized design with her subsidiary parts concealed within the frame, but it should be remembered that all these locomotives are objects of the early machine age. They are, essentially, assemblies of standard parts put together for a certain purpose. All that locomotive building required was such undifferentiated materials as boiler plate, tubing, nuts, bolts and rivets and, in a few cases, some small castings.

In a constructive sense such machine building was primitive, but it permitted adaptability and flexibility to a remarkable extent. It was the American type locomotive that chose to exploit these qualities, finding for them a flambuoyant expression that is a continuing source of delight.

In its earlier days the American type locomotive was decorated with polished brass, shining Russian iron or German silver and eye-popping colors, and each crew took great pride in seeing that its charge was spotless. Even after locomotives became more functional in design and less gaudy in appearance the 4-4-0 retained its dignity because it was so well proportioned.

Perfecting the American type locomotive represented the most valuable engineering work performed on railroad motive power. The work of our most famous and able designers all contributed to perfecting the type, each adding a small part to the finished machine.

We find the first groping towards a locomotive machine was a portable boiler with various accessories attached, such as cylinders and wheels. Then came an arrangement of rectangular beams forming a frame which carried the boiler and provided for the holding of its four-wheel lead truck that carried the whole combination of power generating and transmitting appliances. The clumsy outside wooden beams that acted as frames on the earlier types was abandoned for iron bars that were not susceptible to changes of temperature and the elements, with small extra weight, and to which all operating mechanism was strongly fastened.

The elementary locomotive, consisting of four wheels, two of which were drivers, was deficient in adhesion. Then all four wheels were coupled and this answered the adhesion problem.

With the advent of the lead swiveling truck, one pair of drivers was abandoned for the leading truck and it was soon found again that a single pair of driving wheels made a very slippery engine. Then, just when he was needed, along came Henry R. Campbell who added another pair of driving wheels, the same pair that had been thrown out by the Jervis truck, and won fame and fortune by the invention.

We have already mentioned the very decided improvement on the Campbell, which was introduced the very next year, in the *Hercules* with its equalizing beams between the driving wheels.

Then Thomas Rogers of Paterson, New Jersey, built his first locomotive the same year that the *Hercules* appeared and made his locomotives famous by applying weights in the driving wheels to counterbalance the crank. Coleman Sellers had done this as early as 1835 on two locomotives he built too cheaply for the Philadelphia Railroad, causing him to go out of business and the valuable improvement would have been lost for a time had not Rogers seen its value. Until about 1840 builders of locomotives seemed contented with their work if their engines would haul trains with certainty and regularity—and their products were homely indeed! After 1840 several builders began improving the appearance of their charges, to provide comfort for the engineers and firemen and to build in conveniences for handling. The cab became a recognized necessity and parts that required attention on the road were made easy to reach, while special attention was given to making removal for repairs simple.

For years there was a strong tendency among designers and builders to use the boiler for attachment of frames and even the pedestals that held or braced both moving and fixed parts. By degrees this was found to be bad practice, but they did use the smoke box as a foundation for the cylinders. A point halfway up the smoke box made a handy location for a heavy cylinder so it became a fashion that persisted until a few good designers proved the arrangement to be wrong from an engineering standpoint and moved the cylinders down to the level of the driving wheels.

Wilson Eddy of Springfield, Mass., presented an excellent object lesson when he built his first engine, the *Addison Gilmore*, which had horizontal cylinders and harmonious appearing outlines. Eddy realized that the frame to which was secured the driving wheels made the proper fastenings for the cylinders. His engine represented a sudden jump from the antique to the modern with its level cylinders, divided frames, enclosed cab, cylinder lubrication and sand that could be applied from the cab. Eddy's first engine was built in 1851. To his credit, very few improvements were necessary to produce the perfected American type engine.

For the first forty years of railroad operating, the dominating aim of designers and locomotive builders was to produce an engine suitable for all kinds of service, one that would be fairly efficient and durable enough to make long mileage with small expense for repairs. Except for a few railroads handling minerals and other heavy freight over steep grades the eight-wheel American 4-4-0 type came to be regarded as the ideal engine for handling both passenger and freight trains. Rugged in construction, the American locomotive was simplicity personified. For high speed passenger trains it was fitted with high drivers and for freight use and the long pull the 4-4-0 was built with smaller-diameter wheels.

The man who did so much with and for the American type engine was William Mason. Mason, a textile manufacturer of Taunton, Massachusetts, saw, in October of 1851, a speed and pulling contest by several locomotives on the Western Railroad between Wilmington and Lowell, Mass. The romance of the rails took a quick hold and Mason decided then and there to enter the locomotive building business.

Locomotive builders entered the business to make money and there was small pretense at producing something better or

HEAVY DUTY MOUNTAIN FREIGHT ENGINE

LACKAWANNA & WESTERN RAILROAD

Builder . . . Ross Winans, Baltimore . . . 1854

more artistic than had been built before. The idea was to build a fairly durable engine for the work required, that was as good as the next builder's engine. This was the industry until William Mason entered the business and took up where Wilson Eddy had left off.

Mason loved to build locomotives. He claimed that it took the profits of his textile mills to make up the losses he incurred building locomotives but build them he did. Up to the time Mason began building locomotives in 1852, the idea of art harmony as applied to locomotive design appear to have had no place in the men carrying on the work. They evolved proportions that provided the required strength without using unnecessary material but no apparent attention was bestowed upon the outward appearance of locomotives as far as making the visible outlines harmonious. There was a great deal of ornamentation but the effect was often grotesque where beauty was aimed at. Elaboration of brass in bands and coverings of domes, sand boxes, wheel covers, steam chests and cylinders, with great vagaries of paint on other parts, conveyed the impression of an overdressed woman. These were the days of the red smoke stack and vermilion painted wheels. It was then considered correct to spend hundreds of dollars on the painting of portraits or picturesque scenes on headlights, cab panels and tenders.

Mason took the lead in making locomotives that were handsome without ostentatious paintings. He was a wonderfully ingenious man with a fine artistic sense and his work was always most exquisitely designed. His locomotives were certainly melodies cast and wrought in metal.

His first design was the *James Guthrie*, a 4-4-0 type built for the Jeffersonville & Indianapolis Railroad and named after the road president. It rolled out of his shops at Taunton October 11, 1852. This first engine was no outstanding beauty but led the way to better design with many innovations that were quickly used by other builders. Mason said that most locomotives looked like grasshoppers with their cylinders set above the truck so he spread the pilot axles, lowered the cylinders reasonably close to the rails and placed them in a horizontal position as Wilson Eddy had done. He made the cylinder heads hollow to provide an air space and polished them to prevent the radiation of heat. And the cylinders were interchangable. He made the driving wheels with hollow spokes and hollow rim and put the counter-balance inside the rim of the driving-wheels by pouring hot lead into them. He wanted his front truck wheels to look like the drivers so he made them spoke-type and never put a solid plate wheel on any of his engines, claiming that previously used plate wheels looked like cheeses!

Mason seemed to bring order to heretofore unrelated parts and became known as the "father of the American type engine." He did use color on his engines but was judicious in its use. The wheels were painted a variety of colors, the boilers were almost always covered with Russian iron and the cabs were made of walnut or mahogany wood, highly varnished, and the interior of the cab was well upholstered. Another deft touch was a piece of plate glass fitted over the engine number. Mason's engines greatly helped to establish the form of the American 4-4-0 type, and all that later builders did for many years was increase the proportions.

*   *   *   *

CHAPTER XIV

THE GLORIOUS IRON HORSE

Among the thousands of mechanical devices which we find about us there is none that is more fascinating than the steam locomotive. Not even aviation with its supersonic jets can compare glamorwise with the steam locomotive. Its rugged outlines, its great size, its tremendous power and its aggressive throbbing and ear pounding noises—it has become the perfect symbol of the machine age. When steam locomotion disappears completely from our railroads we are going to lose a part of Americana that can never be replaced but which will, instead, leave a void.

The steam locomotive is a factory in itself, a self contained industrial plant. It has a boiler which turns water into steam; it has an engine in which the expansive power of steam is converted into motion, and it has the machinery which gives motion to the engine. The many forms that locomotive designers and builders evolved in putting these mechanical parts together were, and still are, most stimulating and amazing in their variety.

To amplify our point we have included in our book the *Tiger* locomotive, probably the most colorful and flambuoyant engine ever to steam down the rails. While we cannot vouch for her colors, since they were taken from the builders catalog of the period, and the artists of the day were not above stretching a shade here or there to sell their product (the pink springs, for instance), the locomotive would have looked good in somber black.

One of four 4-4-0 American types built by Matthias Baldwin in 1856-57 for the newly born Pennsylvania Railroad, the *Tiger* certainly has the appearance of having been built by a watch-maker and yet exemplifies the reckless abandonment with which accessory parts such as the steam dome and sand dome was bolted to the boiler, ending somehow in a rakish and fascinating hunk of machinery.

Although the *Tiger* shows that Baldwin still clung to the outside frame in building his locomotives the frame itself had grown so small in becoming flexible enough to adjust to variations in the track and roadbed that it is barely noticeable. However, the *Tiger* was one of the last Baldwin engines to wear a frame outside the wheels. As locomotives grew in size, weight and complexity it became necessary to tuck all structural framework and bracing between the wheels.

Another interesting American 4-4-0 type depicted in these pages is a prime example of tucking gear away and out of sight for even her cylinders and running gear were hidden from view, being set inside, above and between the lead truck wheels! Between 1852 and 1854 the B & O designed eight inside connected 4-4-0 engines which became known as "Dutch Wagons," so named because they looked "foreign" to the accepted appearance of most U.S. designed engines and, because the absence of visible working gear left only the wheels in motion, thus the "wagon" portion of the nickname was born. Another interesting feature of the "Dutch Wagon" was the unusual pilot truck, the only equalizing being the laminated spring fastened to the axle boxes.

Four of these engines were built in the Mt. Clare shops and four were farmed out. Depicted is No. 207 built by Murray & Hazelhurst in 1852. Typical of the cab details of its day about all that was seen inside the cab was a throttle, reverse bar, steam gauge, fire door, tallow cups, bell and whistle cords, lever for the application of sand from the sand dome atop the boiler, and seats for the engineer and fireman.

Relatively few inside cylinder locomotives were designed and built in America in early days, although the type was common in England and on the Continent until recent years. The chief reason the design lacked favor in this country was its inaccessibility of working parts, and the weakness of the then wrought iron cranked axles which the locomotives used to convert the push-pull action of its cylinder pistons into rotary action onto the wheels.

In their heyday these engines must have been beauties. The

as the railroads spread out into new territory where coal was scarce or non-existent and wood abundant the locomotives were designed to burn wood. Then, too, engines that had been designed to burn coal often burned out their grates and flues with the hotter flame generated by coal so could use only wood fires.

Wood was cheap and it burned well, but the exhaust smoke of a wood burning locomotive contained a hail of sparks and burning bits of wood that often set fire to wayside structures and forests and even, on occasion, the train being pulled by the engine. So the smokestacks came to be huge in size to con-

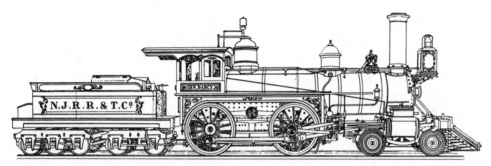

American type Passenger Locomotive built by the New Jersey R. R. & Transportation Company to designs of John Headen, Master of Machinery, the Jersey City Shops, 1867.

outside surfaces of the frame were painted black, the inside vermilion. Headlight, cab, boiler, tender body and splashers were Indian red, the pilot, and all wheels, vermilion. Smokebox and stack were black with a red stripe around the stack's top. Safety valve casing, sand dome, steam dome and the handrails were polished brass. Striping on the wheels was gold as was the lettering and lining on the tender, and the lower edge of the tender had a vermilion stripe.

The big smokestacks that are associated with the American type locomotive did not, as we know, originate with this type engine. They were "Made in U.S.A." products, however. Most of the earliest locomotives had been designed to burn coal but

tain spark-catching devices inside and screens on the tops. They assumed the form of funnels, diamonds and later balloon or "cabbage stacks" as they were called, but they all had one thing in common. Each was designed to arrest sparks and because the wood needed a tall and robust stack to provide the draft for combustion.

By the time railroads reached West Virginia, Kentucky, Indiana and Illinois, coal had been found there in abundant quantity and had been ever abundant in western Pennsylvania so there was coal at both ends of the line, so to speak. More and more railroads began to use this much hotter burning material. When the railroads headed west of the Mississippi they had

MOGUL FREIGHT TYPE

1ST HARD COAL BURNER ON LACKAWANNA & WESTERN R.R.

Builder . . . Danforth, Cooke & Co. . . . 1854

to again fall back on wood, the only available combustible material.

When using coal there was no need for the huge stacks so short smoke stacks appeared on "coal roads." The short stacks, much narrower than their predecessors, were soon nicknamed "cannon" or "shotgun" stacks because of their appearance and because of their sharp exhaust crack, which was no longer muffled inside huge capacity stacks.

The *William Mason* (illustrated), 4-4-0, built by the gentleman of the same name in 1856 at his Taunton, Mass., shops, is a fine example of a huge smoke stacked wood burning locomotive. It is a classic American type engine—and a prime example of the utter disregard for streamlining, an art which was very little understood and totally non-existant in pioneer locomotive building. How much wind resistance the huge smoke stack of the *William Mason* and other engines must have exerted, especially when encountering strong headwinds! Certainly the early engine designers would have been aghast if they had

figures available on the total wind drag offered by their over-size headlights, huge smokestacks, the domes atop their boilers and, last but not least, the squarish and large cabs!

Built for the Baltimore & Ohio Railroad, the *William Mason* was named by that road in appreciation for the previously purchased Mason-built engines that had been giving such excellent service. Many of the parts on the more than 700 locomotives built in the William Mason shops were interchangeable, quite a unique feature for those days, and the B & O and other railroads saved many a dollar thusly. It was only natural that Mason's engines came to be widely copied because of their fine mechanical and performance characteristics, plus the inherent beauty of their styling.

The Baltimore & Ohio used the well built *William Mason* for many years to haul her crack "varnish" (passenger coaches) before putting her to rest in their museum. Carefully preserved, the *Mason* can still travel under her own steam and do it well, having appeared in several movies.

EARLY SMOKESTACK TYPES

The two smokestacks below were used on coal-burning engines, all others mostly on wood-burners.

Balloon or Funnel | Diamond | Large Diamond | Sun-Flower | Rushton or Cabbage Head | Congdon | Capstack | Straight Shotgun or Cannon

THE W^{M.} MASON

FAST PASSENGER ENGINE—B. & O. R.R.

Builder . . . Wm. Mason & Co. . . . 1856

The *William Mason* was a beautiful sight indeed as she steamed down the rails in all her glory.

1ST FORNEY ENGINE

SUBURBAN TYPE DESIGNED BY MATTIAS N. FORNEY

Builder . . . Illinois Central Shops . . . 1862

CHAPTER XV
CIVIL WAR AND RAILS WEST

Much has been written about the part railroads played in the Civil War, the world's first war in which rails moved whole armies, complete to cavalry horses and cannon, supplies and ammunition, and played such a decisive role in the outcome of an entire war.

Of all the railroads involved the Baltimore & Ohio probably suffered greater conflict than any. People who lived along her right-of-way and who worked for her were of divided loyalty. Families and even whole towns split up as folks declared their heartfelt support for one side or the other. And it was a war in which many hundred locomotives and thousands of rail cars were either outrightly destroyed by both North and South troops to prevent their falling into enemy hands or were so ill-used in service that they could never be used again.

Parts of railroads, along with the B & O, belonged to first the North and then the South as the battles raged. Rails were torn up, heated over a bonfire of ties then bent or even wrapped around trees. The Confederates, always short of engines, had an unpleasant habit of helping themselves to motive power. Once they took fourteen of the B & O's best engines from the roundhouse at Martinsburg, West Virginia, and hauled them, forty horses to an engine, down the turnpike all the way south to Strasburg, Virginia, 32 miles away, where they were set on rails again.

By 1863 the Northern troops had conquered enough of the B & O right-of-way and had established the fighting front some miles southward so traffic on the line rose to flood proportions as troop and supply trains by the hundreds had to be moved. Thatcher Perkins, master of machinery, and his fellows kept the Mount Clare shops humming 24 hours a day—forges pounding, lathes turning and shining new locomotives rolling steadily out the doors.

In that feverish year of 1863 Perkins designed and built eleven 10-wheel 4-6-0 type engines to replace the earlier American types for duty in passenger service over the heavy grades of the main line between Piedmont and Grafton, West Virginia. They were the first 10-wheel type passenger locomotives to run on the B & O.

When the locomotives were constructed they embodied the latest developments in design. All the engines had bar frames and cylindrical steam boxes and sand boxes (domes) set on saddles, the latter being a then recent innovation of William Mason, the famous locomotive designer.

Number 117, the *Thatcher Perkins*, first of this new 4-6-0 class, was named by the B & O in honor of Perkins for his latest contribution to engine development. Considered one of the most graceful and well proportioned 4-6-0 engines to run on an American railroad, the *Thatcher Perkins* had drive-wheels of 64½ inches diameter, the others 60-inch except numbers 9, 13, 36 and 136 which wore wheels of 58-inch diameter. Flanges were applied to the rear drive-wheels only on all the engines, to facilitate negotion of sharp curves then existing on the main line. Like other locomotives of the period number 117's cab fittings included only the throttle, reverse lever, try cocks, water-level glass, whistle lever, bell cord, steam gauge, tallow cups and pot, and valve handles to regulate the flow of water and steam from the siphon ejectors.

The *Thatcher Perkins*, along with her sister locomotives, was retired for a time as being too heavy for the war-weary track but later, with new track, pounded the B & O rails for many years. The B & O always thought a lot of her class so never scrapped 117, the one with the largest drive-wheels. For years she rusted in the old roundhouse at Martinsburg then was rejuvenated in 1927 and rests proudly today in the B & O Museum at Baltimore.

In several ways the *Thatcher Perkins* was to signify the end of an era, for the cry was constant for more powerful and thus larger engines. Larger engines meant larger boilers so the smokestacks began to grow smaller for there was a limit to clearance distances at underpass bridges and through tunnels. Also, as man learned how to handle coal there was no need for a spark catching smokestack of huge size, so the larger boilered engines began to swallow their stacks. The use of smaller and yet smaller smokestacks was the first big change to alter the appearance of later engines and then boilers eventually became so large that the once prominent domes began to recede in height for clearance sake and were stretched out to increase their capacity.

Before 1850 designers had used full play of their imagination, but certain standards were gradually evolved. It became unnecessary to build unique and usually one-of-a-kind locomotives, for their designers begans to realize they must design for the traffic conditions encountered and since these conditions varied, two general classes of locomotives were required: Those suitable for hauling freight requiring a high tractive force and comparatively slow speed with small drive-wheels, and those for fast passenger work where tractive force when running could be rather low while the speed must be high with their high drive-wheels, so the builders began making look-alike locomotives by the half-dozen and dozen.

Horsepower in locomotives is the product of tractive force and speed combined so that frequently it was found necessary for passenger locomotives to develop more horsepower than the freight engines even though the two types exerted widely different tractive forces. Since boiler capacity limits horse-power it quickly followed that in proportion to the tractive forces exerted the passenger engine needed larger boilers than did freight engines. The main features required for sufficient boiler capacity were large fire boxes with ample grate area and a large amount of heating surface. The *Thatcher Perkins* of Civil War days is a classic example of the transition to larger boilered passenger locomotives.

Long before the Civil War, Congress had been belabored with requests for a rail route through to the west coast to speed lucrative oriental trade goods to the East. Discovery of gold in California increased the pressure.

Advent of the Civil War made the need of rails connecting both east and west reaches of our country so imperative that in July 1862 Congress passed the Pacific Railroad Act, authorizing establishment of the Union Pacific Railroad Company, granting a right-of-way 200 feet wide through public land on each side of the roadbed and 100 feet wide on either side of the roadbed through private property. The Act also granted ten alternate sections per mile of public domain on both sides.

The Union Pacific, chartered in 1862, began construction westward at Omaha, Nebraska, in 1864. The Central Pacific, also authorized to build eastward from the west coast to meet the U. P., had already been incorporated in 1861. Construction on the Central Pacific began in 1863.

Both railroads were granted subsidies of $16,000 a mile in territory which governmental surveys deemed level, $48,000 a mile for construction through mountains, and $32,000 a mile for track laid between mountain ranges. Thus the two companies began their now historic construction race to see which could pile up the greatest amount of subsidies. Locomotives for use on the Central Pacific Railroad were shipped around the Horn by boat from their eastern builders and most of the engines used by western roads continued to be built by companies east of the Mississippi, throughout the history of steam locomotion in the U. S.

C. L. FLINT

3 Ft. Gauge Double-Truck Engine

Builder . . . Wm. Mason Machine Works . . . 1876

CAMEL TYPE FREIGHT ENGINE

DESIGNED TO BURN WASTE ANTRACITE COAL

Builder . . . Philadelphia & Reading Shops, John E. Wooten, Supt. of Mach. . . . 1877

CHAPTER XVI

CYLINDERS, BOILERS, COAL AND HUGE ENGINES

Soon after the steam locomotive was fairly well perfected in the United States there was a mania towards building locomotives capable of doing at least 60 miles an hour, even when the track was not safe at half that speed. They built the high speed high-drivered engines and then learned that the boilers would not generate steam fast enough to keep the wheels turning. Associated with big driving wheels were cylinders so large they gulped the available steam in short order so the designers and builders soon learned the intimate relation between the heating area of the boiler and the cylinder size. Large cylinders also caused excessive drive-wheel slippage when the weight thereon was too little; so through expensive trial and error the proper proportions for a good working locomotive were found.

For years the length of stroke of the piston within the cylinder was a matter of much debate. When the *Stevens*, with her huge 8-foot diameter wheels was designed the cylinder size settled on was 13-inch diameter and a 3-foot stroke! The first engine built of that type was, of course, found deficient in tractive power so they compounded their error on the succeeding *Stevens* class engines by increasing the piston stroke to 38 inches! That stroke was the longest ever tried for locomotives. In two engines with a single pair of driving wheels, built by William Norris in 1850 for the Erie Railroad the cylinders were 14"x32". The experience gained on these engines indicated to designers that a shorter stroke with a larger base cylinder would produce better results. Gradually a cylinder with the diameter about 7/10ths the length of the stroke became practically standard.

The boiler is the real measure of the capacity of a locomotive but the cylinders are the measure of power and the proportions of heating surface and grate area should be based on the capability for generating the volume of steam required by the all-important cylinders.

For several decades a practice prevailed on American railroads which indicated that certain officials thought that by judicious tampering, an engine could be made to do work far beyond its designed capacity. A locomotive with certain size cylinders was found to be capable of hauling, say, 500 tons over the limiting grade of a division. The officials knew enough about cylinder capacity to figure that an increase of 1 inch in the diameter would increase the tractive power about 15 percent and enable the engine to haul 75 more tons, a sizable amount. Increasing the cylinder capacity without any other change proved disappointing as the engine proved less efficient than before. It was the ancient attempt of trying to force a quart into a pint bottle.

As late as the year 1870 the argument as to the relative merits of different locomotive types had not been resolved, nor had the lessons of standardizing been learned by some railroads, the Erie Railroad being a prime example. Their locomotives had been built by nearly every builder in the country and the company had 83 different patterns of locomotives! There was a belief among railroad men that certain makes of locomotives were better than other makes and that they could do more work without exact regard to their dimensions. Trainmen and others became strong adherents of certain makes of engines and bigoted enough to quarrel blindly in favor of their pets.

Later engineering proved that one engine must exert as much tractive power as another, dimensions and steam pressure being equal. There were differences in valve gear that would make one engine smarter in starting, while a second would be more persistent in pulling on long grades.

There was also for many years quite an argument stirring as ot the relatives merits of wood vs. coal. Coal had wisely been tried on early American railroads in an effort to get a hotter fire to the locomotive boilers to make the engine do more work through its cylinders, but it was too hot to handle as it burned the grates and other parts away. Coal was dirty and wood was cheap and in most areas plentiful. When suppliers of wood began price squeezing or when certain areas could not supply wood at any price, coal was once again experimented with and found to be at least 48 percent cheaper than wood for the heat value received. Many trials took place on nearly every railroad and a fire resistant brick bridge or "arch" above the hottest blast was found to be the best answer. This method of lining the furnace has held all through the days of steam.

As anthracite coal was found to be an ideal fuel for locomotives, being cheap in the eastern states and emitting far less smoke than wood, it was only natural that attempts were made from the beginning of railroads in the U.S. to utilize it. Yet, twenty years after the first locomotive operated in the United States, using coal as fuel, wood was the fuel used by locomotives, even when their principal job was hauling coal to market!

Peter Cooper had burned coal in his *Tom Thumb* and the B & O had used anthracite with some success in their early vertical-boilered locomotives. Early experimenters with coal burning locomotives on the B & O and other railroads had the theory that concentration was necessary to maintain a very hot fire, so limited their grate area. After long years of failure the locomotive builders discovered that anthracite, being a slow burning hard coal, needed a vastly larger grate area than wood or bituminous soft coal to produce an equal amount of heat.

When this lesson was learned coal began supplanting wood when the former was readily available and the locomotives themselves had to undergo many changes to conform to the new fuel, mostly by growing larger.

The favorite 4-4-0 American locomotive eventually reached the limits of its capacity and the increasing use of coal helped antedate the type. The grate area limits and the steam producing power of the boiler restricted the capacity of this marvelous design. In 1870 probably 85 percent of all locomotives at work in the U.S. were American type and while heavier, more powerful engine types took over the increasingly heavier freight duties it was the fancy, gaudy and fleet American 4-4-0 that proudly reigned as queen of the passenger runs on many railroads for twenty-five more years.

For over 20 years the American locomotive was the Rome towards which nearly all designers traveled. Then passenger cars became larger and more luxurious and much heavier. Freighting demands increased steadily too, so railroad managers commanded that more powerful locomotives be provided. The "American" type locomotive reached its zenith in 1872. In that year the Baldwin Locomotive Works built 422 engines, the average weight in working order, 64,000 lbs. and most of the engines were of the 4-4-0 type.

The intensity of the popular desire to keep that type of locomotive in use was seen in the many ingenious efforts made to enlarge the grate area. The first movement was to increase the distance between the driving wheels so the grates could be lengthened. Side rods as long as 9 feet came into use, but the increase in grate area proved a poor remedy. Then came the practice of sloping the grate and raising the center line of the boiler. By doing this the back of the grate was brought high enough to pass over the rear axle, permitting the fire box to extend an infinite distance. This permitted the grate to be made as long as it could be fired! Such grates were sometimes made

POWERFUL FREIGHT ENGINE

DESIGNED BY A. J. STEVENS, MASTER MECHANIC

Builder . . . Central Pacific Railroad Shops, Sacramento . . . 1882

from 9 feet to 10 feet long and so was always unpopular with the poor fireman. It was also wasteful of coal.

Enough improvements were made by different designers and builders that prolonged the utility of the 4-4-0 engine so that it continued to haul the plush name passenger trains until 1896 and even longer on a few railroads. In 1891 William Buchanan of the New York Central R.R., co-operating with the Schenectady Locomotive Works, brought out an abnormally large 4-4-0 locomotive to haul the heavy express trains. It was numbered 870, had cylinders 19″x24″, driving wheels 78″ in diameter, weighed 120,000 lbs., of which 80,000 lbs. were upon the drivers.

That form of engine was largely copied and made even heavier, one group of engines being made in which the engine weighed 136,000 lbs. with 90,000 lbs. on the drivers. This was exceeding the limit, for 22,500 lbs. weight pressing the rail beneath each wheel was more than the steel rails or steel tires of its day could endure from a fast running locomotive. Much greater weight per wheel eventually came to be used but the rails were made much heavier and stronger, too.

Long before the "American" locomotive had reached its safe weight for heavy fast express trains, railroad companies had proceeded to build specific engines for either passenger or freight service. The mountain and mineral roads had developed heavy motive power and ten-wheelers, mogul 2-6-0 and consolidation 2-8-0 engines were familiar to all railroads. About 1885 the eight-wheel engine was rapidly disappearing from the front of freight trains, even on level roads, and its place taken by the above mentioned types, the ten-wheel engines decidedly in favor because they closely resembled the favorite 4-4-0. In many quarters there was a decided prejudice against the mogul on the ground that a pony (two-wheel) lead truck was not so safe in leading the engine as a four-wheel truck. That was contrary to the teachings of experience and of engineering principles. Most ten-wheel engines had their front pair of drivers so far forward that there was little weight on the four-wheel lead truck and there were cases where the hind drivers dropped into a low track joint and the back of the engine jerked downward far enough to lift the lightly loaded front truck off the track. There were many mysterious derailments of the ten-wheel locomotives that were probably caused by this fact.

In 1877 John E. Wootten introduced the wide fire box extending over the frames and which provided all the grate area any locomotive might require. Wootten's invention improved on forms previously designed by Colburn and Milholland; but as general manager of the Philadelphia & Reading Railroad he had the means of pushing the merits of his boiler. The Wootten fire box was used principally for burning fine anthracite; but it was beginning to find favor with railroads using bituminous coal when, in 1895, a new rival appeared that made rapid progress into popularity.

It was a new engine type known as the *Atlantic*, a 4-4-2 engine designed by William P. Hensey of the Baldwin Locomotive Works. This engine design permitted the fire box to extend over the small trailing wheels, enabling the grate area to be greatly increased, doubled if necessary. This form of locomotive proved to be the most popular later type produced. It rapidly pushed the 4-4-0 form, long so popular, out of demand and found favor in all countries where powerful, fast locomotives were required.

Three years before Baldwin's brought out the first *Atlantic* engine, they built another locomotive, also designed by Mr. Hensley, which was known as the *Columbia* 2-4-2 type. That engine permitted the fire box to be extended over the rear wheels, just as the *Atlantic* does, providing unlimited grate area, as was later done on the *Atlantic*. Somehow the railroad world did not take kindly to the *Columbia*, probably through the same prejudice that prevailed against running moguls on fast passenger trains. Yet, a later engine, called the *Prairie*, a 2-6-2

FAST EXPRESS LOCOMOTIVE

New York Central & Hudson River Railroad

Builder ... Schenectady Locomotive Works ... 1891

151

type designed by Waldo H. Marshall in 1901 for the Lake Shore and Michigan Southern Railway, became highly popular as ideal motive power for heavy express trains. No fault was found with the 2-6-2 engines, but the 2-6-0 mogul was not in demand!

The way was paved for the design and building of really powerful king size freight engines when, in 1876, Anatole Mallet, a French engineer, designed some two-cylinder compound locomotives for a French railroad. In this type engine the two cylinders were placed in the usual manner, one on each side of and below the fire box but one cylinder was large and the other smaller. Steam was fed to the small cylinder, then after this cylinder had used the steam to push its piston, the remaining exhaust steam, being lower in pressure now, was reheated and used in the larger cylinder on the other side of the firebox. The first engines were small but were very successful and paved the way for the extensive use of compound locomotives which was to come later.

Mallet was ambitious to develop the compound system for unusually powerful locomotives so he set about designing a four-cylinder compound type which was to have two sets of drive-wheels, one set of wheels for a high pressure cylinder on each side of the locomotive and one set of low pressure cylinders on each side. The high pressure cylinders were fastened to the frames in the usual position, one on each side of and below the fire box and the low pressure cylinders were carried on a truck actuating a second set of driving wheels aft of the front set. This last feature, placing cylinders and driving wheels into a swiveling truck instead of being affixed rigidly to the frame was called articulating, thus we have, with the two features mentioned above, the articulated compound Mallet engine type. Mons. Mallet built his first of this type engine about 1888 and was the only man to ever have a distinct locomotive type named after him.

Three-cylinder compound engines were tried later, the third cylinder being placed directly under the fire box and between the two outside cylinders, the third cylinder using exhaust steam from the two outside cylinders. The inside cylinder drive rod was connected to the front axle which was cranked. It was the four-cylinder compound principle, however, which was developed into the huge powerful articulated engine of this 20th century, used all over the world but reaching gigantic proportions in the United States.

It is curious the stupendous results that sometimes spring from small causes. As cities became larger and spread out and smaller cities and towns sprang up outside the metropolitan areas there was a need for a suburban type engine, one that could run equally well in either direction and haul commuters to and from the downtown areas of the big cities. Some of this early traffic was being handled by horse-drawn cars and a lot of people were commuting by horse and buggy then just as folks do today by automobile. Then, in the summer of 1872, there was an epidemic of distemper among horses in all large cities in America which highly stimulated the need and demand for suburban railroads that could be operated by locomotives.

When the workaday dependability, speed and efficiency of the suburban locomotive was repeatedly demonstrated the elevated railroads in Chicago and New York resulted. Several prominent and a few small companies came to the aid of suburban needs of the locomotive—Matthias N. Forney, Baldwin, H. K. Porter Co., Vulcan Iron Works and the Davenport Locomotive Works supplying the majority of locomotives.

CHAPTER XVII

EULOGY OF STEAM

The American locomotive of the late 1850's was a real Jim Dandy—full rigged and high sporting. Men spoke of them with the same high emotion with which they discussed race horses or opera stars or the big figures in the still-new national game of baseball. Small boys adored them. With all this emotion there was also a bit of real reverence. The claims of rival types of locomotives were discussed by their adherents with vehemence and even violence.

Wrote the great Elihu Burrit of the locomotives of that day and generation, "I love to see one of these high creatures, with sinews of brass and muscles of iron, strut forth from his stable and salute the train of cars with a dozen sonorous puffs from his iron nostrils, then fall back gently into his harness. There he stands, champering and foaming upon the iron track, his great heart a furnace of glowing coals, his lymphatic blood boiling within his veins; the strength of a thousand horses is nerving his sinews; he pants to be gone."

It was an age of elegance and of rich magnificence which was reflected in the glorious American locomotive of that day . . . in its masses of shining brass wrought in some parts into intricate and delicate scroll design, in the gay paintings, landscapes and portrait and patriotic emblems of every sort, upon the great square headlights, the ornate cabs and on the sides of tenders . . . in the gay colors of the engines themselves, reds, greens and yellows, and always a profusion of gold gilt stripings and letterings. All the really great engines were named, and in their very names there was amazing variety and beauty.

It soon came to pass that the larger railroads began experiencing trouble in keeping the names of their locomotives straight and found it hard to constantly be entering names in their various ledgers. So, in the interest of efficiency they began numbering their engines, retaining, in some instances, the names also. In 1862 a man named Cornelius Vanderbilt appeared on the U.S. railroad scene, a man destined to change greatly the appearance of locomotives. Having started earlier with a lone New York ferry boat, Vanderbilt had the true Midas touch that turned everything he handled on sea or land, into gold. Busily engaged in buying and consolidating railroads into a virtual empire, now the New York Central Railroad, he paused for a moment in 1870 and took a long look at his gaily painted and shining locomotives. Vanderbilt noted that it was the duty if not the pride of the crew to carefully polish their engines and figured that thousands of pounds of cotton waste and hundreds of manhours were annually wasted in the process and quickly decreed that all his engines be painted . . . black! Locomotives were still growing ever larger so, in the interest of economy, most American railroads followed by adopting this practice. No doubt the poor engine crew and the roundhouse painters and wipers had had a time of it keeping their engines in anything approaching their original beauty but many crewmen and railfans the country over must have felt the effort well spent.

Other railroads followed Vanderbilt's example by painting their locomotives black while a few retained their gaudy coloring for years after, especially western railroads.

Engine crewmen have always taken great pride in their locomotives. While a relatively few engineers would "beat" their engines by the rough application of throttle and the slamming of the reverse lever to slow or stop the train, most of them took pride in and some felt a deep actual love for their fiery steeds. In the early days the engineers were assigned a particular engine and they nursed these engines with fatherly care. There was one old timer that would beckon his fireman over to his side of the cab, then he would pull out his watch and point at it—"See that watch, bub?" he'd shout. "The only difference between my watch and my engine is size!"

Men like Vanderbilt could decree that their locomotives be painted black but there still remained the handrails, bell and whistle which was of shiny brass. Most engineers managed to get to the roundhouse several hours before their appointed run and would go over their iron steeds with meticulous scrutiny, wiping here, polishing there, checking the fire box to make sure the fire was properly banked, and always carrying an oil can or "tallow pot" to make sure their charges were "lubed" for the journey ahead.

As the fascination of railroading enclosed America in its grasp the locomotive engineer became the nation's new hero, the Brave Engineer, the man on the right hand side of the locomotive cab who drove his train safely through night and storm; or tragically into another train, or washout, or defective bridge or rail. The Brave Engineer largely supplanted the Soldier and the Sailor as what Young America wanted to be when he grew up. Keen-eyed, tanned and wind-burned to leather by the elements, the lines in his face creased with soot, the engineer always looked ahead, scanning the bright rails for danger, his mind weighted with responsibility for his train, his left hand always at the throttle. It was his duty to put her through or die at his post, and die he too often did.

He was a great, a magnificent figure to Young America. More than one banker, college president and successful businessman envied him too, for this was the post they all wanted and once meant to have, the calling for which superb whistles blew and noble bells rang to the melody of pounding drivers on the rails and the singing of the steam exhaust.

And did not this hero of ours share a little bit of his exalted position with us simply by acknowledging our wistful hand wave by returning it with a friendly wave of his own, taking time from his wondrous job just for us!

Time was his very god, this man. No matter what the time card showed, no matter how able the officials, the division superintendents, the dispatchers, switchmen and firemen, it was at last and finally the engineer who put her through "on the advertised." With the coming of the railroad, time for the first moment in history really became an important measure in the lives of most Americans. To be "on time" was "railroad fashion." Even an ordinarily detached man who lived in a cabin in the woods by a quiet pond once reflected on time and the railroad. "They come and go," said he of the trains that passed his pond, "with such regularity and precision, and their whistles can be heard so far that the farmers set their clocks by them, and thus one well regulated institution regulates a whole country. Have not men improved somewhat in punctuality since the steam railroads came? Do they not talk and think faster in the depot than they did in the stage-coach office?"

In 1893 an event transpired that put the capper on the first seven decades of steam locomotion—frosting on the cake that showed how far steam had progressed. The West Albany Shops of the New York Central & Hudson River Railroad had built an American 4-4-0 type locomotive with 86-inch drivers for an assault on the world's speed record. Then, on May 10, 1893, while pulling the Empire State Express train the sleek and somber engine, number 999, attained a speed of 112.5 miles an hour, a record which stood for several years.

And so it came to be that, fueled with nothing more than wood from our forests and raw coal from beneath the earth, our early steam locomotives forged a new country into a closely-knit, free and prosperous nation, criss-crossed and held tightly together with rails of iron.

Steel engraving of the famous "999" which set a speed record of 112.5 mph, May 10, 1893. The locomotive is still preserved as an example of motive power of the late nineties.

SPECIFICATION CHART

SPECIFICATIONS OF PRINCIPLE LOCOMOTIVES DEPICTED

	ENGINE NAME	ENG No.	BUILT 18--	WEIGHT POUNDS	LENGTH FEET	HEIGHT FT. IN.	CYLINDERS INCHES	PRESSURE POUNDS	DIAMETER DRIVERS INCHES	TUBES No.	WHYTE WHEEL	NOTES
1	TOM THUMB *		30	10800	13' 2 11/16"	12' 8 7/8"	5×27 (1)	90	30 1/4	38	4 WHL	TRACTIVE PWR 820 LBS
2	BEST FRIEND	1	31	10000	-	-	6×16	50	54	108	"	" " 400 LBS
3	YORK *		31	13540	10' 5 5/8"	11' 8 3/8"	5 1/4×12	115	30	108	"	" " 935 LBS
4	DEWITT CLINTON * (1)	1	31	6750	11' 6"	-	5 1/2×16		54	50	"	DELIVERED APP. 10 HP
5	OLD IRONSIDES	1	32	11000	-	-	9 1/2×18	-	54	72	"	BOILER 30" DIA
6	MISSISSIPPI		34	14000	-	-	9 1/2×16	-	43	-	"	TRACTIVE PWR 4821 LBS
7	ANDREW JACKSON	7	36	27160	14' 7"	12' 5"	12 1/2×22	75	36	169	"	REWORKED AS ATLANTIC
8	MONSTER		36	60900	-	-	18×30	-	48	-	0-8-0	4 OF THIS TYPE BLT. VARIED
9	LAFAYETTE *	13	37	29200	29'	14'×4 1/2	9×18	90	48	56	4-2-0	TRACTIVE PWR 2323 LBS
10	WINANS "CRAB" TYPE		37	24000	-	-	12 1/2×24	-	36	36	4 WHL	4 OR MORE BUILT
11	ALERT	7	37	24120	24' 5 1/2"	11'3"	10×18	100	54	108	4-2-0	TRACTIVE PWR 3428 LBS
12	GOWAN & MARX		39	22000	-	-	12 1/8×18	-	42	-	4-4-0	10 OTHERS BUILT
13	CUMBERLAND	37	44	47000	-	-	17×24	-	33		8-WHL	MORE THAN 12 BUILT
14	WASHINGTON		47	34675	-	-	13×18	-	46	-	0-6-0	TRANSFERRED TO P.R.R. 1849
15	CHESAPEAKE		47	44000	-	-	14 1/2×22	-	46	-	4-6-0	20 BLT FOR P.R.R.
16	MEMNON	57	48	74700 (A)	30'10"	13'9"	17×22	65	41	120	0-8-0	TRACTIVE PWR 8580 LBS
17	PERRY		48	50975			17×22		43	-	0-8-0	SISTER ENGINE "DAUPHIN"
18	INDIANA		49	47000			14×20		72	-	6-2-0	3 OF THIS TYPE BUILT
19	LIGHTNING		49	40000			16×22		84	-	6-2-0	
20	STEVENS		49	46000			13×38		96	-	6-2-0	7 BLT SPECS VARIED
21	PIONEER P.R.R.		51	25000	26' (B)	11'2"	8 1/2×14	-	54	63	2-2-2	
22	CAMEL (WINANS)		54	55000	-	-	19×22	-	43	-	0-8-0	1ST OF TYPE BLT 1848
23	WILLIAM MASON	25	56	56000	48'4"(B)	14'2"	16×22	75	58	107	4-4-0	TRACTIVE PWR 6225 LBS
24	THATCHER PERKINS	117	63	90700	53'1"(B)	14'2"	19×26	75	58	140	4-6-0	" " 10350 "
25	CAMEL (HAYES)	217	73	77100	30'3 7/8	14'3 1/2	19×22	65	50	124	4-6-0	" " 8775 "
26	J.C. DAVIS	600	75	90600	40' 2 1/2"	14'8 1/2"	19×24	65	56	165	2-6-0	" " 8580 "
27	MODERN AMERICAN TYPE	870	91	120000	57'1 3/4(B)	14'8"	19×24	-	78	269	4-4-0	SEVERAL BUILT

NOTE: WEIGHT AND DIMENSIONS ARE FOR LOCOMOTIVE ONLY UNLESS OTHERWISE INDICATED.

(A) TOTAL LIGHT WEIGHT OF ENGINE AND TENDER.

(B) TOTAL LENGTH, ENGINE AND TENDER.

* REPLICA.

(1) SPECIFICATIONS ARE FOR ORIGINAL LOCOMOTIVE, NOT THE REPLICA DEPICTED

INDEX

Locomotives are indicated by italic type followed by an asterisk.

Addison, Gilmore 134
Alert* 62, 64, 66, 130
Allen, Horatio 15, 18, 32, 65, 72
American type locomotive 71, 72, 74, 77, 133, 134, 136, 137, 138, 140, 144, 148, 150, 153
Andrew Jackson* 53, 56
Arabian* 53
Atlantic* 50, 53, 56
Atlantic 4-4-2 type locomotive 150
Baldwin Locomotive Works 148, 150, 152
Baldwin, Matthias W. 29, 32, 49, 57, 60, 62, 66, 74, 77, 137
Baltimore & Ohio R.R. 9, 10, 15, 23, 24, 50, 53, 56, 60, 65, 66, 71, 77, 78, 137, 140, 143, 144, 148
Beaver Meadow Railroad 66, 72
Bell 65, 72, 137, 143, 154
Best Friend of Charleston* 15, 18, 24, 32, 50
Boston & Worcester Railroad 66
Brakes, train 130
Brooks, James 71
Brother Jonathan* 49
Brown, William H. 26
Buchanan, William 150
Burrit, Elihu 153
Camel* 78
Camden & Amboy Railroad 29, 65, 71, 72, 78
Campbell, Henry R. 71, 134
Canals 9, 61, 62
Central Pacific Railroad 144
Chicago & North Western Railway 64
City Point Railroad 74

Civil War 60, 78, 143, 144
Clark, Mr. 24, 26
Columbia 2-4-2 type locomotive 150
Columbia Railroad 57
Columbus* 56
Comet* 66
Consolidation 2-8-0 type locomotive 150
Cooper, Peter 9, 10, 12, 50, 148
Crab type locomotive 56, 57, 77
Cumberland* 77
Darrell, Nicholas W. 15, 18
Davenport Locomotive Works 152
Davis & Gartner 53
Davis, Phineas 23, 49, 50, 53
Delaware & Hudson Canal Company 9, 49
Detmold, C. E. 15
DeWitt Clinton* 24, 26, 49
Dragon* 77
Dripps, Isaac 65, 72, 78
Eastwick & Harrison 57, 74
Eddy, Wilson 134, 136
E. L. Miller* 32, 49
Empire State Express 154
Engineer, locomotive 15, 18, 20, 23, 49, 53, 60, 62, 66, 78, 129, 130, 134, 138, 154
Erie Canal 61, 62
Erie Railroad 57, 147
Experiment* 49, 52, 57, 71
Fireman, locomotive 18, 53, 62, 66, 78, 80, 129, 130, 134, 138, 148, 154
Flexible beam truck 77

Forney, Matthias N. 152
Galenta & Chicago Union Railroad 62
Garrett & Eastwick 66, 72
Gartner, Israel 23, 50, 53
George Clinton* 53
George Washington* 53, 57, 60
Grasshopper type locomotive 50, 53, 56, 60
Gillingham, George 53, 56
Gowan & Marx* 74
Harrison, Joseph Jr. 72
Haskell, Mr. 29
Hayes, Samuel Jr. 78
Headlight 62, 65, 66, 74, 77, 133, 136, 138, 153
Hensley, William P. 150
Hercules* 72, 74, 134
Horse 9, 10, 12, 15, 23, 24, 26, 29, 53, 57, 61, 62, 71, 143, 152, 153
Indiana* 80
Imlay, Robert 50
James Gunthrie* 136
James, J. H. 66
James Monroe* 53
James Madison* 53
Janney, Eli Hamilton 129
Jeffersonville & Indianapolis Railroad 136
Jervis, John B. 24, 49, 57, 134
John Bull* 29, 65, 72
John Hancock* 53, 56
John Stevens* 78, 80, 147
John Quincy Adams* 53, 56

INDEX

Locomotives are indicated by italic type followed by an asterisk.

Knight, Jonathan 50
*Lafayette** 60
Lake Shore & Michigan Southern Railway 150
Leicester & Swannington Railway 66
Leipzig & Dresden Railroad 56
*Lightning** 80
*Locomotion** 9
Mad River & Lake Erie Railroad 66
Mallet, Anatole 150
Mallet type locomotive 150
Marshall, Waldo H. 150
*Martin Van Buren** 53
Mason, David 29
Mason, William 134, 136, 140, 143
Matthews, David 24, 29
*Memnon** 77
*Mercury** 53, 65
Michigan Central Railroad 62
Miller, E. L. 15, 32
Mogul 2-6-0 type locomotive 150
Mohawk & Hudson Railroad 24, 49
Monster type locomotive 72, 74
Mount Clare Shops 53, 137, 143
Murray & Hazelhurst 137
Museum, Baltimore & Ohio 56, 60, 144
Museum, Henry Ford 26
Museum of Science and Industry
Museum, Peale
Newcastle Manufacturing Company 77
New York & Erie Railroad 78
New York Central & Hudson River Railroad 154
New York Central Railroad 24, 62, 150, 153

Norris, Edward S. 80
Norris, Richard & Son 78
Norris, William 57, 60, 74, 147
*Old Ironsides** 29, 32, 52
Paterson & Hudson River Railroad 66
Pennsylvania, Philadelphia & Reading R.R. 78
Pennsylvania Railroad 134
Perkins, Thatcher 143
Philadelphia & Reading Railroad 29, 74, 150
Philadelphia, Germantown & Norristown Railroad 29, 32, 71
Philadelphia Railroad 134
*Phineas Davis** 53
*Phoenix** 32, 50
*Pioneer** 62, 64
Prairie 2-6-0 type locomotive 150
Planet type locomotive 72
Porter, H. K. Company 152
Rail 9, 15, 20, 24, 32, 49, 50, 53, 56, 57, 60, 62, 64, 71, 72, 74, 77, 129, 130, 143, 150, 154
Reeder, Charles 53
Rogers Company 57
Rogers, Thomas 134
Rogers, Ketcham & Grosvenor 57, 66
*Samuel P. Ingham** 66
*Sandusky** 66
Schenectady Locomotive Works 80, 150
Schuylkill & Susquehanna Calal 61
Sellers, Coleman 134
Smithsonian Institute 9
South Carolina Canal & Railroad Company 15, 18, 24, 32, 65, 72

*South Carolina** 32, 52
Southern Railway 15, 18
South Hadley & Montague Canal 61
Steere, A. G. 57
Stephenson, Robert 9, 10, 18, 65, 72
Stevens, Robert L. 71, 72
Stockton & Stokes 10, 12
*Stourbridge Lion** 9
*Thatcher Perkins** 143, 144
*Thomas Jefferson** 53
*Tiger** 137
*Tom Thumb** 10, 12, 23, 24, 50, 148
Track 9, 10, 12, 18, 20, 26, 29, 50, 57, 65, 66, 71, 72, 77, 78, 129, 150, 153
*Traveller** 53
Trenton Locomotive Works 74
Tyson, Henry 78
Union Pacific Railroad Company 144
Utica & Schenectady Railroad 62, 80
Vanderbilt, Cornelius 153, 154
Western Railroad of Massachusetts 77
Western Railroad 134
Westinghouse, George 130
West Point Foundry 9, 15, 24, 32, 49
*West Point** 18, 24
*William Galloway** 60
Winans, Ross 50, 53, 56, 77, 78, 129
*William Mason** 140
Whistle 65, 66, 72, 130, 137, 143, 154
Wootten, John E. 150
*York** 23, 50

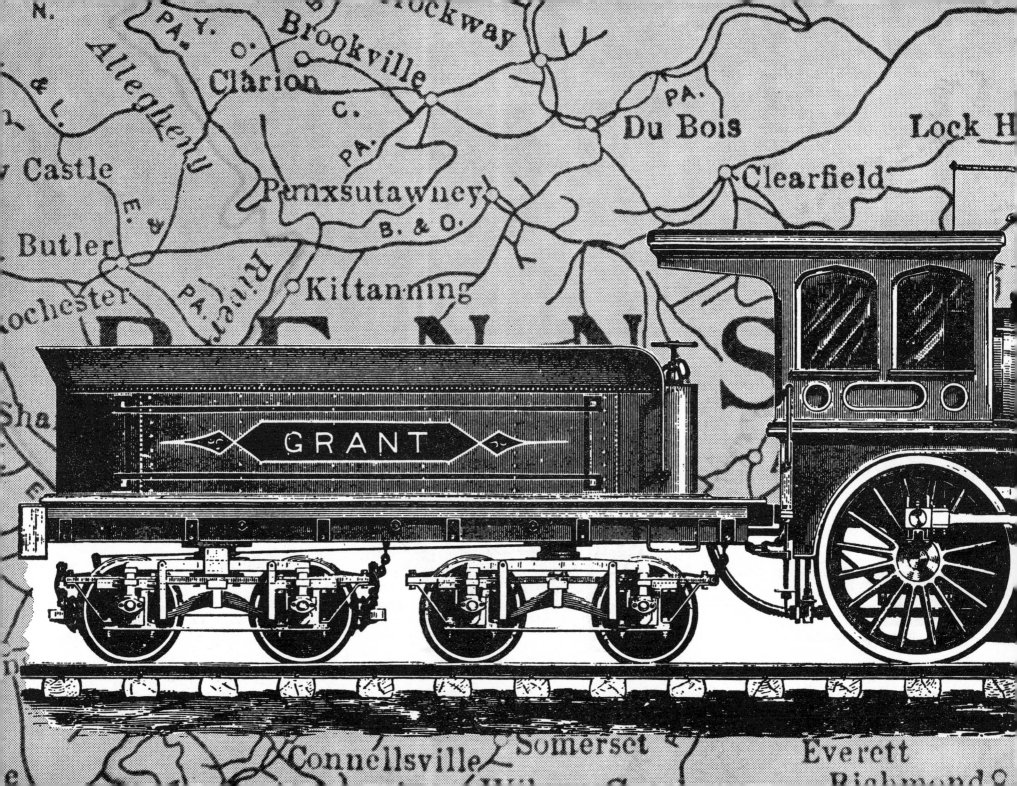